Reflections on Statistics: Learning, Teaching, and Assessment in Grades K–12

D1516821

The Studies in Mathematical Thinking and Learning Series
Alan Schoenfeld, Advisory Editor

Reflections on Statistics: Learning, Teaching, and Assessment in Grades K–12

Edited by

Susanne P. Lajoie
McGill University

LEA LAWRENCE ERLBAUM ASSOCIATES, PUBLISHERS
1998   Mahwah, New Jersey   London

Lawrence Erlbaum Associates, Inc., Publishers
10 Industrial Avenue
Mahwah, New Jersey 07430

Cover Design by Kathryn Houghtaling Lacey

Library of Congress Cataloging-in-Publication-Data

Reflections on statistics : learning, teaching, and assessment in grades K–12 / edited by Susanne P. Lajoie.
      p. cm. — (Studies in mathematical thinking and learning)
    Includes bibliographical references and index.
    ISBN 0-8058-1971-1 (cloth : alk. paper). — ISBN
0-8058-1972-X (paper : alk. paper)
    1. Mathematical statistics—Study and teaching. I.
Lajoie, Susanne. II. Series.
    QA276.18.R44 1998
    519.5'071—dc21   97-7357
                                        CIP

Printed in the United States of America
10 9 8 7 6 5 4 3 2 1

Contents

Preface

Susanne P. Lajoie
McGill University

One call for reform in mathematics education is to address statistics at the precollege level. We created an interdisciplinary working group consisting of mathematics educators, cognitive scientists, teachers, and statisticians to address this call in a manner that could help teachers and researchers make informed decisions about how to introduce statistics in grades K–12. The interdisciplinary nature of this group stimulated a lively interchange of ideas for enhancing the learning, teaching, and assessment of statistical understanding. Mathematical educators contributed their insights into how teachers teach mathematical ideas, and heightened the group's awareness of the ecological needs of the current mathematical classroom. Cognitive scientists shared their understanding of developmental differences in learning and provided theoretical perspectives that contribute to the design of effective learning environments. Classroom teachers shared their ideas about classroom activities and assessment of student learning, as well as their concern for in-service workshops that could help them acquire skills in this new content area. Statisticians shared their concerns about what was feasible to teach in these early grades, and their views of statistical literacy. One of the outcomes of this multidisciplinary collaboration is this volume, which represents a partial consolidation of some of our interactions and, as such, should appeal to both practioners and researchers.

It is too early in our research to provide prescriptions for statistics instruction and assessment in grades K–12, and hence we provide some reflections on statistics, trying to develop a clearer image of where the field may be going in terms of teaching, learning, and assessment of statistical content that is appropriate for different grade levels. In an attempt to identify the important key ideas that need to be considered for children and adolescents learning statistics, the book is organized along four interdependent themes: content, teaching, learning, and assessment. By focusing chapters on particular themes, we intend to cultivate a better understanding of how each area relates to improvements in statistics education by (a) clarifying the important mathematical ideas in statistics; (b) investigating student learning from a cognitive perspective describing how students formulate, represent, reason about, and solve statistics problems, examining the construction of ideas via the social interaction that occurs in and out of school; (c) investigating the beliefs, pedagogy, and content knowledge of teachers; and (d) ensuring that teachers have a deep understanding of statistics so that they can provide a classroom environment that stimulates deep thinking by their students (Romberg, in press).

A premise shared by the contributors to this volume is that if statistics are introduced in the K–12 period, students will be better prepared for decision making in the real world. Our quest is to find ways to provide students with opportunities to do statistics in a manner that prepares them for real-world experiences. Lajoie and Romberg introduce the volume by describing the main issues that have arisen in our working group as they pertain to statistical content, teaching, learning, and assessment. Although each contributor discusses all four areas, their focus for this book is on one theme. Statistical content is addressed by two chapters. Scheaffer, Watkins, and Landwehr's chapter provides a detailed account of what a high-school graduate should know about statistics and at what grade levels certain statistical content should be introduced. They introduce the notion of designing statistical curriculum strands, where statistics content is introduced and extended through critical periods in the curriculum. These periods are elementary, upper elementary, and high school. Burrill and Romberg focus on statistical content for middle school and suggest ways in which this critical period can serve to bridge the gap between informal knowledge about making conjectures from collections of data and the formal ways users of statistics analyze data. The teaching statistics section presents one chapter on teaching statistics to upper elementary students (Bright and Friel) and one chapter on teaching statistics to teachers through a professional development workshop (Friel and Bright). The uniqueness of these two chapters is that they demonstrate the needed links in pedagogy between understanding the needs of the learner and of the teacher, with particular emphasis on demonstrating activities that extend the knowledge of both parties. The three chapters in the learning section cover students' learning statistics and probability from primary school to adulthood. The chapters by Horvath and Lehrer (2nd graders, 4th and 5th graders, adults) and Metz (primary grades) carefully examine the dimensions of students' intuitions and the transitions in this knowledge that reflect emerging statistical competence. Derry, Levin, Osana, and Jones focus their discussion of statistical learning in terms of statistical gaming situations where students engage in simulations of professional activities such as conducting research, preparing presentations, and participating in hearings and conferences in which students are required to develop statistical arguments centering on their own statistical evidence. Assessing statistics is covered in three chapters that explicitly address how to integrate assessment with instruction. Lajoie, Lavigne, Munsie, and Wilkie, and Schwartz, Goldman, Vye, Barron, and the Cognition and Technology Group describe ways in which technology can be used to enhance the assessment process through modeling student performance so that students can internalize the assessment goals of instruction. Lajoie et al. describe new assessment methods for monitoring student progress in statistics in a grade-eight classroom. Schwartz et al. discuss ways of stimulating learning through the direct confrontation of competing prototheories. Gal describes a multilayered approach to the assessment of the reasonableness of student opinions about statistics. Finally, Lajoie provides an epilogue that presents her reflections on what can be learned from this volume and some suggestions for future work in this area.

This volume represents an interdisciplinary effort to construct an understanding of how to enhance statistics education and assessment for students in elementary and secondary schools. The audience that would appreciate this book includes:

- teachers interested in learning about research on statistical problem solving that has been conducted in classrooms;
- mathematical educators interested in the contributions from cognitive scientists regarding how students best learn statistics;
- cognitive scientists and educational psychologists interested in how mathematical educators have addressed statistical content at various grade levels by involving teachers in workshops that focus on the learning process;
- statistics educators who are interested in research that can assist them in their own teaching and assessment practices.

ACKNOWLEDGMENTS

A number of people and organizations are responsible for making the publication of this book possible. The writing of the individual chapters and the preparation of this book were made possible through funding from the United States Office of Educational Research and Improvement, National Center for Research in Mathematical Sciences Education (NCRMSE). Partial support was provided by the Social Science and Humanities Research Council in Canada. I would like personally to acknowledge Thomas Romberg for his foresight and insights in this area, for encouraging me to pursue this research focus, and for convincing me that this book was indeed timely. I gratefully acknowledge the contributions of the members of the NCRMSE Statistics Working Group (Gail Burrill, Sharon Derry, Susan Friel, Iddo Gal, Joan Garfield, Chris Hancock, Jeff Horvath, Victoria Jacobs, Cliff Konald, James Landwehr, Nancy Lavigne,

Richard Lehrer, Joel Levin, Susan Mayhew, Kathleen Metz, Steven Munsie, Helen Osana, Thomas Romberg, Andee Rubin, Leona Schauble, Richard Scheaffer, Dan Schwartz, Ronald Serlin, Michael Shaughnessy, and Tara Wilkie) for their help in formulating many of these ideas. I would like to express my gratitude to Alan Schoenfeld, our series editor, for his insights and contributions to the editorial process. I further acknowledge the assistance of my graduate assistant, Nancy Lavigne, for her tireless efforts in assisting me with the day to day responsibilities of managing this book.

REFERENCES

National Council of Teachers of Mathematics. (1989). *Curriculum and evaluation standards for school mathematics.* Reston, VA: Author.

Romberg, T. A. (in press). The social organization of research programs for the National Center for Research in Mathematical Sciences Education. Madison: National Center for Research in the United States.

Identifying an Agenda for Statistics Instruction and Assessment in K–12

Susanne P. Lajoie
McGill University

Thomas A. Romberg
University of Wisconsin-Madison

Reflections on Statistics: Learning, Teaching, and Assessment in grades K–12 was written with the intent of opening a dialogue on a topic that is still in its infancy. Statistics have traditionally been reserved for university-level courses and consequently little information (from research or practical experience) exists to guide the implementation of the National Council of Teachers of Mathematics' (NCTM) *Curriculum and Evaluation Standards* (1989) vis-à-vis statistics at the elementary and secondary levels. For this reason, a statistics working group was established by the National Center for Research in Mathematical Sciences Education (NCRMSE) to try to breach this chasm in the literature. This group has formally met and engaged in ongoing dialogues for several years. The composition of the group is international and includes members from the International Study Group for Research on Learning Probability and Statistics. The composition of the NCRMSE working group accurately reflects the interdisciplinary nature of the field. Teachers, mathematics educators, cognitive scientists, statisticians, and educational psychologists have shared their knowledge in an attempt to clarify what we understand statistics instruction and assessment to mean for precollege grade levels. This edited volume attempts to provide some insights for new researchers and educators in this field.

BACKGROUND

One template for developing statistics instruction at precollege levels is to examine the literature on statistics education at the university level. The widespread consensus among educators and researchers is that statistics education, in its current form, is inadequate (American Statistical Association, 1991; Mosteller, 1988; NCTM, 1989; Posten, 1981; Shaughnessy, 1992). A discouraging finding regarding statistics is that most individuals have a limited understanding of the subject matter. Given that statistics instruction has generally been restricted to those pursuing professional or academic careers, this limited understanding is not surprising. What is discouraging, however, is the finding that a large proportion of university students fail to understand elementary statistics concepts (Garfield & Ahlgren, 1988), even after taking several courses (Posten, 1981). Students' lack of conceptual understanding may be due, in part, to their having had insufficient early exposure to statistics instruction (Posten, 1981; Shaughnessy, 1992). This perception has led to the consideration of teaching statistics in the early grades. Because models of statistics instruction at the university level are quite sophisticated, and not always effective, new approaches for teaching and assessing statistics in K–12 must be developed. Statistics instruction must be redesigned to include activities that empower the student, are engaging, provide students with opportunities for *doing* statistics, and demonstrate relevance to real-world applications (Lajoie, 1995; Lajoie, Jacobs, & Lavigne, 1995; Lajoie, Lawless, Lavigne, & Munsie, 1993). A problem-solving focus can help promote statistical understanding by providing students with opportunities to inquire, investigate, analyze, and interpret rather than to compute and memorize (Lajoie, in press-a).

A synopsis of the "big issues" identified in each section of this volume is provided here as a sneak preview and as a possible agenda for further statistics research.

STATISTICAL CONTENT FOR SCHOOL MATHEMATICS

Statistics, according to the NCTM (1989) guidelines, can be viewed as a branch of mathematics dealing with the collection, analysis, interpretation, and presentation of sets of data, where a statistic is a quantity or form of representation based on quantities derived from those sets of data. The problem that the content subgroup addressed was how to operationalize some of these reforms into content that would be appropriate at each of these grade levels. Their first question concerned the "big ideas" that reflect this branch of mathematics and the proper sequence for fostering the development of such ideas in children and adolescents.

The content subgroup struggled with its selection of statistical content as well as with the appropriate sequence of instruction. However, the group's members have made great headway in defining the scope of such content (see Scheaffer, Watkins, & Landwehr, this volume) as well as in providing concrete examples from an implementation of such content in the middle school (see Burrill & Romberg, this volume). Scheaffer et al. concluded that not enough is known about how statistical reasoning can be progressively realized through the appropriate developmental timing of introducing instruction involving statistical situations (problem situations might come from statistics in literature or history or statistics found in newspapers and magazines). However, a joint committee of the National Council of Teachers of Mathematics and the American Statistical Association (NCTM-ASA) has been searching for such situations in real classrooms in order to inform educators about the successes in practice. One possibility that is proposed in this volume is to have a statistics content theme carried across grades so that the appropriate connections between contexts could be made. Burrill and Romberg provide an exemplar of how a statistics and probability strand could be implemented in the middle school. Their examples from the *Mathematics in Context* units (NCRMSE & Freudathal Institute, in press) provide a convincing picture that statistical understanding is possible at these grade levels.

Statistical content and the learner's conceptual understanding of the content need to be investigated concurrently. One of the problems in describing statistical content at the precollege level is that there is no precedent to help us determine what to teach. The developmental literature would suggest that "readiness" should be considered when students are developmentally ready to learn statistical concepts (Lajoie, Jacobs, & Lavigne, 1995). Others suggest that informal knowledge is quite strong in the early grade levels and can be used to guide statistics instruction (Fischbein, 1975; Konold, 1991). What is still missing is the cognitive research that documents how this informal knowledge can be used to guide instruction. At the elementary level, several people have started to address these issues in statistics by examining children's intuitions about modeling (see Horvath & Lehrer; Metz, this volume). Each of these chapters presents concrete statistics tasks that promote statistical understanding. Only by documenting cognitive skills and the characteristics that differentiate the skilled from the unskilled learner for each content area (i.e., learning about measures of central tendency or covariance) can we improve the teaching and assessment of such understanding.

We must communicate the statistical content we expect students to acquire at different grade levels. Scheaffer et al. provide a good idea of what high-school students are capable of when they handle such content, and how the appropriate prerequisite knowledge should be taught so that students learn to connect their new knowledge with old knowledge. These researchers have started to address what is needed to promote statistical understanding (by addressing the statistical content needed to promote the construction of statistical relationships) and what is needed to extend statistical knowledge (by applying knowledge in new situations; Carpenter & Lehrer, in press; Lajoie, in press-b). Identifying the structural relationships within statistical content is a key step to facilitating the instruction and assessment process. The *Assessment Standards for Teaching Mathematics* (NCTM, 1995) is quite clear in its guidelines for tying instructional goals to assessment goals. By making assessment goals open to the learner, the learner has a greater chance of assessing his or her own progress at attaining such goals. Instruction and assessment of probability would look quite different from the instruction and assessment of inferential statistics, and such instruction would vary across grade levels. In defining what we expect students to acquire in the K–12 grade levels, we communicate what statistics we find meaningful for students to learn. We must take into consideration what is meaningful to children in the context of their everyday lives and how this meaningfulness changes from grade to grade.

At a general level, one content theme that appears throughout this volume is that students learn to "critique" as well as to "produce" statistics (Burrill & Romberg; Derry, Levin, Osana, & Jones; Schwartz, Goldman, Vye, Barron, & The Cognitive Technology Group at Vanderbilt; Gal; Lajoie, Lavigne, Munsie, & Wilkie; Scheaffer et al.). Such an approach to instruction is paralleled in fostering reading comprehension, where students learn to critique reading summaries as well as to produce their own summaries (Palinscar & Brown, 1984). In statistics, we are concerned that students learn to interpret and evaluate the statistics they read or hear about, as well as to generate their own statistical investigations when required (Gal, this volume). By encouraging students to develop a critical attitude toward the role of statistics in contemporary society, it is possible to increase our overall statistical literacy (Gal, in press; Joram, Resnick, & Gabriele, 1995). Such an attitude would imply going beyond the use of specific techniques taught in statistics courses. Students must become consumers as well as producers of statistics. Scheaffer et al. go as far as stating that "every high-school graduate must be educated to be an intelligent consumer of data and to know enough about the production of data at least to judge the value of data produced by others." It is just as important to make intelligent use of information presented statistically as to do the analysis oneself. Becoming a good consumer of statistics implies developing a critical attitude that allows one to raise questions about problem situations. A major goal should be that students learn to build a coherent case to answer a question. Hypothesis generation is a critical aspect of statistical reasoning not commonly addressed in statistics instruction. Students need to learn how to pose questions and how to gather data that will build a case for their question. They should learn to argue and reason about the representativeness of the data they collect, organize their data appropriately, and present their data in graphs and tables. Computers can be used as tools for assisting students in their case building, but students still need to learn to defend their statistical reasoning.

At a very basic level, students must be given content that supplies them with statistical and mathematical techniques to help them make sense of the data. These techniques include procedures for organizing, visualizing, and summarizing their data. Both descriptive and inferential statistics must be taught. Students should be able to make predictions and decisions based on data, and probability should be taught hand in hand with sampling and design. However, Scheaffer et al. (this volume) clearly state that statistics is not about numbers, but about numbers in context.

STATISTICAL PEDAGOGY

Once we identify what to teach, decisions must be made regarding how to teach it. However, statistics teaching must consider both the learner and the teacher. From the learner's perspective, we know that good statistics pedagogy must provide a proper foundation for students in the early grades, one that includes making sense of data and feeling comfortable with what must be done with the data in later grades. In other words, teaching must build on students' prior knowledge so that students can construct the appropriate relationships with what they know before they extend this knowledge and apply it to new situations. Bright and Friel (this volume) address these issues in the context of teaching students about graphical representations and interpreting data. They give clear examples of how to teach students to construct relationships and how to extend students' understanding through multiple contexts. Schwartz et al. (this volume) similarly provide opportunities for the extension of statistical knowledge through a macrocontext, a problem-solving situation anchored through a videodisk that situates the learner in the context of an extended problem, which can be worked on for several days.

Part of the difficulty with statistics pedagogy in K–12 is that the content is as new for many teachers as it is for students. Consequently, teachers must be empowered in their teaching through appropriate workshops where they learn new techniques for teaching new content. Teachers must be provided with appropriate preservice and inservice training that will give them the knowledge base they need to feel comfortable teaching about data and chance. Friel and Bright (this volume) have taken the lead in designing such training. Teachers should be given the same opportunities as students to hear and model intellectual discourse about statistics. Teachers need release time and support to learn new skills such as how to assess children's thinking in statistics. Teachers need to know and understand possible intervention strategies for dealing with student errors and misconceptions. Time and classroom management of data collection and analysis must also be considered. For instance, should twice as much time be devoted to analysis and discussion as to data collection? Teachers must be able to resolve divergent points of view and synthesize material for the class, and they can only do so given proper training.

At the moment, statistics instruction is considered within the mathematics curriculum and, thus, can be integrated into other areas of mathematics such as algebra, geometry, and computations such as fractions and percents. However, statistical concepts can also be used as a springboard into other disciplines and for applications outside the classroom. It is possible to make statistics more meaningful for the learner by making these connections across subject matter more clear. Clarifying these connections to students demonstrates that statistics can be applied to any domain, thereby promoting the notion of statistical literacy. When teaching statistics, activities may be designed that integrate statistics across disciplines and demonstrate the usefulness of statistics within the real world. Derry, Levin, Osana, and Jones (this volume) have demonstrated how statistics can be incorporated in science and social studies classroom instruction by having students evaluate statistical evidence in these domains and prepare debates on controversial issues pertinent to such subjects. Lajoie, Lavigne, Munsie, and Wilkie (this volume) have provided opportunities for eighth-grade students to apply their understanding of statistics to research questions that students themselves generate and apply to real-world concerns.

Instructional interventions can improve the frequency and quality of statistical reasoning. Classrooms that promote statistical understanding generally provide the appropriate tasks, tools, and normative practices (Carpenter & Lehrer, in press; Lajoie, in press-b). Statistical tasks are described in each chapter, but highlighted in the statistical content section. Many of the statistical tasks are problem-based and cooperative in nature. A central theme in many of the pedagogical approaches discussed here is that statistics is introduced to students in the context of a question, where students discuss statistics in an open-ended problem. Furthermore, a variety of experiences are provided so that students can integrate different representations of information in multiple contexts.

Technology is discussed throughout the volume as a tool for restructuring instruction in a way that fosters the real-world application of theories learned in schools (Lajoie, 1994). New technologies are changing the ways in which problem situations and methods of representation are used in instruction (Romberg, in press). Technology, as illustrated in chapters by Lajoie et al. and Schwartz et al., can be considered as a cognitive tool (Lajoie, 1993; Perkins, 1985; Salomon, Perkins, & Globerson, 1991) designed to enhance learning. Through these tools, students can practice knowledge in the context of the overall process of statistical investigation rather than practice statistics skills in isolation, such as performing simple computations. The computer supports lower-level cognitive skills, thereby enabling learners to utilize the remaining resources for higher-order thinking. Furthermore, software applications designed to foster graphing and data analysis assist students in generating and testing hypotheses in the context of their problem solving. The technological support is often in the form of exemplifying students' thinking through visual traces of their problem-solving ability and through modeling the appropriate components of statistical problem solving (see in this volume, Lajoie et al.; Schwartz et al.) with concrete examples of other students solving similiar problems. In this manner, data collection, data representation, interpretation, and argumentation can all be modeled through the use of technology.

Communities of learning can be developed whereby statistical problem solving and reasoning become part of the normal classroom practice. Good pedagogy fosters such practice where classroom dialogues can be utilized by the teacher to engage all students in reflecting on the importance of the way data are collected, graphed, and analyzed (see in this volume, Derry et al.; Bright & Friel). Sharing group projects in the classroom is another mechanism for fostering a community of statistical problem solvers (see Lajoie et al., this volume).

LEARNING STATISTICS

Researchers in the learning subgroup discuss how their research can shed light on how students learn statistics. Understanding emerges from mental activities that involve constructing relationships, extending and applying knowledge, reflecting, articulating, and making knowledge one's own (Carpenter & Lehrer, in press). Although some chapters may focus on one dimension more than another, these dimensions of understanding are reflected in each section of this book. Two of the chapters represented in the learning section (Horvath & Lehrer; Metz) provide detailed examinations of how children develop an understanding of chance and uncertainty. They provide clear evidence of transitions in mental models of statistical understanding and, in so doing, start identifying signposts or benchmarks of statistical understanding within the realm of chance and uncertainty. Derry et al. focus their discussion of statistical learning in terms of statistical gaming situations where students engage in simulations of professional activities such as conducting research, preparing presentations, and participating in hearings and conferences in which students are required to develop statistical arguments centering on their own statistical evidence. These professional activities are paralleled in Lajoie et al.'s chapter, where students engage in the statistical design process as they develop their own statistics projects. Lajoie et al. describe naturally occurring situations where students raise questions about the real world, follow through with their investigations, and defend their findings to their peers. Similar learning situations occur in Schwartz et al.'s work where students anchor their learning in a technology-based macrocontext. Two questions that are touched on in this section are: (a) What is the role of different representational systems, particularly computer-based dynamic representations, in the development of understanding? and (b) What are the effects of different types of mentoring on student thinking? There is a real need for longitudinal research in this area because learning transitions in such authentic situations may require time.

ASSESSMENT OF STATISTICAL UNDERSTANDING

Statistical understanding is difficult to define and assess in general terms. Students must be given opportunities to communicate their statistical knowledge regardless of the statistical content, especially the processes by which they reach an answer. Often this communication must be verbal rather than written. Verbalizations may be more useful than written text for certain age groups and may, in fact, lead to richer articulations of what students know or do not know. For instance, in a middle-school study, Jacobs (1993) found the quality of students' written work inferior to their verbal reasoning of statistics and probability. This finding supports the need for multiple methods of assessment (Collins, Hawkins, & Frederiksen, 1994; NCTM, 1995) whereby all children are given a fair medium in which to provide evidence of their learning. When every individual is provided with ample opportunity to demonstrate his or her understanding in a manner that is conducive to his or her strengths, the equity standard described in the *Assessment Standards* is truly met. Thus, multiple assessment methods are more equitable and valid than a single assessment measure. Chapters by Gal, Lajoie et al., and Schwartz et al. describe these issues more fully. In each of these chapters there is evidence that students' omission of statistical content is not necessarily an indication that students lack understanding but that they may simply have neglected to articulate their understanding of that content. For this reason, the authors contend that developing scripted questions for assessing students' performance on verbal tasks is crucial. Analyzing the answers to such questions through verbal protocol analyses can lead to insights regarding the development of students' statistical reasoning.

Classrooms that promote understanding tend to marry instructional and assessment goals. These goals are facilitated by providing opportunities within instructional practice for student reflection and articulation of knowledge. Many of the chapters in this volume provide such opportunities (see Derry et al.; Gal; Lajoie et al.; Schwartz et al.). Cognitive research, designed to identify what it means to be proficient in statistical problem solving, can be used to improve instruction and assessment. Once the components of statistical competence are identified, they can be modeled for the learner based on the level of the learner's competency. If a statistical misconception is assessed, it must be remedied through appropriate instruction because it will not "disappear" by itself (Fischbein, 1975).

There are several statistical problem-solving components that should be considered in modeling and assessing statistical competence. As with other problem-solving tasks, the way students represent the problem will determine how well they solve it. A major portion of problem representation rests with the ability to understand the problem statement. For example, how do students interpret statistical vocabulary? Words that reflect "probability" to a statistician, may simply reflect "outcomes" to students (Konold, 1989, 1991).

Another major part of the task may be formulating the research question and following through with data creation, collection, and interpretation. Lajoie et al. speak to these components in their research by providing exemplars of the statistical problem-solving process to assist the novice learner. Lehrer and Romberg (in press) and Hancock, Kaput, and Goldsmith (1992) pay particular attention to the data-creation phase of problem solving, concentrating on the process of planning data collection based on the expected data analysis. Special attention must also be paid to the representation of the data once it is collected. For instance, do students understand how to organize the data in a manner that allows them to conduct the appropriate data analysis for their research question? Selecting the appropriate type of analysis and designing a study to support this analysis are major components of statistical problem solving. Once students perform their analyses, do they understand their results?

As research grows in the area of statistics instruction and assessment for precollege students, it will be necessary to build a model of what statistical competence means in the context of different content areas. Only then can we improve the quality of instruction and assessment for students at different levels of learning. These models, once established, can be used to empower students in the use of statistics. Future research in the area of precollege statistics should include an integrated framework for the instructional and assessment process. Such a framework would include looking at the statistical content, the learner's conceptual understanding of the content, and the ways instruction builds on the assessment of the learner in the context of an instructional situation. This volume attempts to shed light on some of these issues.

ACKNOWLEDGMENTS

Preparation of this document was made possible through funding from the Office of Educational Research and Improvement, National Center for Research in Mathematical Sciences Education (NCRMSE). I gratefully acknowledge the assistance of the NCRMSE Statistics Working Group for their help in formulating many of these ideas.

REFERENCES

American Statistical Association. (1991). *Guidelines for the teaching of statistics K–12 mathematics curriculum.* Landover, MD: Corporate Press.

Carpenter, T. P., & Lehrer, R. (in press). Learning mathematics with understanding. In E. Fennema & T. A. Romberg (Eds.), *Classrooms that promote understanding.* Mahwah, NJ: Lawrence Erlbaum Associates.

Collins, A., Hawkins, J., & Frederiksen, J. R. (1994). Three different views of students: The role of technology in assessing student performance. *Journal of the Learning Sciences, 3* (2), 205–217.

Fischbein, E. (1975). *The intuitive sources of probabilistic thinking in children* (C. A. Sherrard, Trans.). Boston: D. Reidel.

Gal, I. (in press). Statistical tools and statistical literacy: The case of the average. *Teaching Statistics.*

Garfield, J., & Ahlgren, A. (1988). Difficulties in learning basic concepts in probability and statistics: Implications for research. *Journal for Research in Mathematics Education, 19*(1), 44–63.

Hancock, C., Kaput, J. J., & Goldsmith, L. T. (1992). Authentic inquiry with data: Critical barriers to classroom implementation. *Educational Psychologist, 27,* 337–364.

Jacobs, V. R. (1993). *Stochastics in middle school: An exploration of students' informal knowledge.* Unpublished master's thesis, University of Wisconsin, Madison, Wisconsin.

Joram, E., Resnick, L. B., & Gabriele, A. J. (1995). Numeracy as cultural practice: An examination of numbers in magazines for children, teenagers, and adults. *Journal for Research in Mathematics Education, 26* (4), 346–361.

Konold, C. (1989). Informal conceptions of probability. *Cognition and Instruction, 6* (1), 59–98.

Konold, C. (1991). Understanding students' beliefs about probability. In E. von Glasersfeld (Ed.), *Radical constructivism in mathematics education* (pp. 139–156). The Netherlands: Kluwer.

Lajoie, S. P. (in press-a). Authentic statistics: Authentic for whom? *Proceedings of the American Statistical Association meeting.*

Lajoie, S. P. (in press-b) Understanding of statistics. In E. Fennema & T. A. Romberg (Eds.), *Classrooms that promote understanding.* Mahwah, NJ: Lawrence Erlbaum Associates.

Lajoie, S. P. (1995). A framework for authentic assessment in mathematics. In T. A. Romberg (Ed.), *Reform in school mathematics and authentic assessment* (pp. 19–37). Albany: SUNY Press.

Lajoie, S. P. (1994). Technologies for extending learning. In P. Brusilovsky, S. Dikareva, J. Greer, & V. Petrushin (Eds.), *Proceedings for the East-West Conference on Computer Technologies in Education, Part 1* (pp. 16–21). Yalta, Crimea: EW-ED.

Lajoie, S. P. (1993). Computer environments as cognitive tools for enhancing learning. In S. P. Lajoie & S. J. Derry (Eds.), *Computers as cognitive tools* (pp. 261–288). Hillsdale, NJ: Lawrence Erlbaum Associates.

Lajoie, S. P., Jacobs, V. R., & Lavigne, N. C. (1995). Empowering children in the use of statistics. *Journal of Mathematical Behavior, 14* (4), 401–425.

Lajoie, S. P., Lawless, J., Lavigne, N. C., & Munsie, S. D. (1993, April). *New ways to measure skills of problem solving, reasoning, communication, and connectedness.* Paper presented at the annual meeting of the American Educational Research Association, Atlanta, GA.

Lehrer, R., & Romberg, T. A. (in press). Exploring children's data modeling. *Cognition and Instruction.*

Mosteller, F. (1988). Broadening the scope of statistics and statistical education. *The American Statistician, 42* (2), 93–99.

National Center for Research in Mathematical Sciences Education and Freudenthal Institute. (in press). *Mathematics in context: A reformed curriculum for grades 5–8.* Chicago: Encyclopedia Brittanica Educational Corporation.

National Council of Teachers of Mathematics. (1995). *Assessment standards for teaching mathematics.* Reston, VA: Author.

National Council of Teachers of Mathematics. (1989). *Curriculum and evaluation standards for school mathematics.* Reston, VA: Author.

Palinscar, A. S., & Brown, A. (1984). Reciprocal teaching of comprehension-fostering and comprehension-monitoring activities. *Cognition and Instruction, 1,* 117–175.

Perkins, D. N. (1985). The fingertip effect: How information processing technology shapes thinking. *Educational Researcher, 14,* 11–17.

Posten, H. O. (1981). Review of statistical teaching materials for 11–16-year olds. *The American Statistician, 35* (4), 258–259.

Romberg, T. A. (in press). *The social organization of research programs for the National Center for Research in Mathematical Sciences Education.* Madison, WI: National Center for Research in the United States.

Salomon, G., Perkins, D. N., & Globerson, T. (1991). Partners in cognition: Extending human intelligence with intelligent technologies. *Educational Researcher, 20,* 10–16.

Shaughnessy, J. M. (1992). Research in probability and statistics: Reflections and directions. In D. Grouws (Ed.), *Handbook for research in mathematics teaching and learning* (pp. 465–494). New York: Macmillan.

PART ONE

STATISTICAL CONTENT

CHAPTER ONE

What Every High-School Graduate Should Know About Statistics

Richard L. Scheaffer
University of Florida, Gainesville

Ann E. Watkins
California State University, Northridge

James M. Landwehr
Bells Labs—Lucent Technologies, Murray Hill, NJ

In 1993, many Americans were shocked to hear news reports of a survey that found that "twenty-two percent of Americans doubt that the Holocaust ever occurred." How could this be? How could so many doubt the occurrence of one of the most significant events of the 20th century? The answer lies in the poll, or survey, itself and in the difficulty of communicating with language, which always produces some degree of imprecision. The question asked by the Roper organization was worded as follows.

> Does it seem possible or does it seem impossible to you that the Nazi extermination of the Jews never happened?

Among the respondents, 22% said "it seemed possible" and another 12% said they did not know. Only 65% said it was "impossible it never happened." The double negative in the question, it seems, confused the respondents. Picking up on this point, the Gallup organization conducted a follow-up poll that asked the question more clearly.

> The term Holocaust usually refers to the killing of millions of Jews in Nazi death camps during World War II. In your opinion, did the Holocaust: definitely happen, probably happen, probably not happen, or definitely not happen?

Among the respondents to this poll, 83% said the Holocaust definitely happened and another 13% said it probably happened. Only 1% said it definitely did not happen. Quite a difference! Gallup also asked the Roper question to another group of people and found that 37% of the respondents said it seemed possible that the Holocaust never happened.

Surveys (polls) and experiments increasingly guide political, medical, and business decisions, but they are little understood. An excellent study of the trends in collecting and using data is the book *Tainted Truth* by Cynthia Crossen (1994), which pointed out that

> We are skeptical about statistical and factual information, but not as skeptical as we think. . . . We respect numbers and we cannot help believing them. Yet, more and more of the information we use to buy, elect, advise, acquit and heal has been created not to expand our knowledge but to sell a product or advance a cause. . . . That's what surveys do, they basically manufacture news.

People depend on data to make intelligent decisions, yet the data they see are often tainted. So, what can be done? Part of the answer lies in education. Every high-school graduate must be educated to be an intelligent consumer of data and to know enough about the production of data to form reasonable judgments about the value of data provided by others. High-school graduates should understand how surveys and experiments work, how good surveys and experiments can be designed, and how data can be properly analyzed. This education must be built into the K–12 curriculum, primarily in mathematics but with support and applications from the sciences, the social sciences, health, and other academic subjects.

This chapter describes the statistical content that should be part of the K–12 curriculum to achieve these goals. The topics are organized and discussed in terms of five broad strands: number sense, planning a study and producing data, data analysis, probability, and statistical or inferential reasoning. Although this order of the topics is in roughly increasing level of sophistication and intellectual difficulty, successful education requires treating each of these topics in several different contexts and over a range of grade levels. Specific grade-level recommendations for individual topics are not given in this chapter. The remainder of this introduction provides a brief overview of the chapter's sections.

The first section discusses how "number sense" can be developed in students. Beginning in the elementary grades, students should observe the world around them and correctly use the numbers they see every day within the context of their own lives. Students begin to collect data by sorting, counting, and measuring items of practical use or interest to them. As number sense develops, students can begin to correctly make and interpret graphs and tables, which are mainstays of the information society in which they will live.

With initial experiences such as these, students are on the road toward understanding how data is produced in planned studies to answer specific questions. By the upper elementary grades, experiments and surveys should be part of classroom activities that help solidify concepts of number and data. In the middle- and high-school years, experiments and surveys should be repeated but with increasingly sophisticated topics appropriate for those ages. Key concepts for a sample survey, for an experiment, and for planning and critiquing studies are given in the second section.

After a study is designed around a specific question or problem and the data is collected, analysis of the data can begin. The third section offers examples and discusses key concepts of data analysis.

Probability is a notoriously difficult subject for students, and over the years probability topics have been taught more than statistical topics. The fourth section discusses which probability concepts are most important and how they can be taught, as well as which probability topics can be deemphasized for the purpose of producing quantitatively literate citizens. In brief, probability as a relative frequency needs to be introduced in the elementary grades. Probability must be revisited and the notions expanded to that of probability distributions in the middle and higher grades. Simulation is the key tool for developing and understanding the topics.

High-school students should be expected to have a full range of data-analytic skills and be comfortable designing an appropriate study for a question of interest to them, collecting the data, and summarizing the results. They should also understand key probability concepts. The fifth section describes how, at this point, statistical (or inferential) reasoning can be developed. Teaching a wide variety of inferential techniques is neither necessary nor desirable. Rather, students should understand key basic concepts of statistical reasoning so that they can begin to judge the merits of quantitative arguments in the world around them.

NUMBER SENSE

Jim Swift, a master teacher of statistics, often remarks that students should be encouraged to be "professional noticers" of the numerical world around them. He encourages them to formulate their observations as specific questions: The temperature seems cooler today; is it actually cooler? The sign says 10% off; off what? Are the best CD players always the most expensive ones? How much do I really have to earn in order to buy a car? How many minutes per week do I actually exercise? Did four-laning that street improve traffic flow?

Number sense begins by seeing numbers in context—data—and then making judgments about the data through comparisons with other data from the same or similar contexts. Here is the digit 4. What does it mean? It is a mathematical abstraction until it can be placed within a context. If the unit "pounds" is added, then the number begins to take on meaning. If the context is a discussion of the size of a trout recently caught, the number takes on a direct and favorable meaning. If the context is about a newly born baby, then the number takes on an unfavorable meaning, because a 4-pound baby is destined to be in some difficulty. If the context is pressure in a water line, one may not have a judgment about the number because of lack of experience with other numbers from this context.

Understanding Variables

The concept of a variable helps students to organize data into meaningful groups so that useful summaries and comparisons can be made. In this age of graphing calculators and computers, most students find that a spreadsheet organization of the data is helpful. An example follows.

Data are collected from each student in a class on the number of compact discs owned and the typical amount of spending money per week. The gender of the person responding is also recorded. A *spreadsheet organization* means that the data can be arranged in tabular form, as shown in Table 1.1 for responses from four students. The variables are the column headings, and the values that the variables assume are listed underneath, one value per variable for each student in the study. The units attached to the values should be clearly stated; in this example CDs is in number per student and spending is in dollars per student. The variable Gender does not have numerical values; it is a categorical variable.

Interpreting Tables and Graphs

Original data or summary values often are placed in a table or on a graph to show patterns and make comparisons. Understanding variables, as discussed earlier, helps a student understand tables and graphs constructed by others or determine the best table or graph to use in constructing an original display. Understanding the data to understanding the display is clearly a two-way street.

TABLE 1.1
Data in Tabular Form

Gender	CDs	Spending
F	11	5.00
M	3	21.50
F	42	15.00
F	0	18.75

Data can be one of three types: categorical, count, or measurement. A poll of students on a political issue usually involves placing each student's opinion into a category. A study of number of television sets per house generates a count on each house in the investigation. An experiment to evaluate a new diet involves a measurement of weight loss (literally on a scale) for each person on the diet. Categorical data are summarized by frequencies, whereas counts and measurements are summarized by center (What constitutes a typical value in the data set?) and variability (What constitutes a typical distance from the center of the data set?).

The appropriate graph for a set of data is determined by the type of data. If the data are categorical, then the only meaningful numbers to graph are the frequencies or percentages of responses in the various categories, which is nicely handled by a bar graph. If, on the other hand, the data are counts or measurements, then they should be plotted in some manner that keeps track of both their value and frequency. A number-line plot, stem-and-leaf plot (stem plot), or histogram can serve this purpose well. Students should learn to look for appropriate measures of center and variability, symmetry versus skewness, clusters and gaps, and possible outliers. The shape of the data distribution aids in determining the types of summary measures that are useful.

Tables are arrays of numbers in rows and columns. The value at the intersection of a row and column, called *joint* information, is defined in terms of both the row and the column in which it occurs. Again, this value may be raw data or a summary value. (In most tables, it is more likely to be a summary value.) If data are summarized for each column (across the rows) the summary becomes *marginal* information on the column headings; if data are summarized for each row (across the columns) the summary becomes marginal information on the row headings. By concentrating on a single row or column, to the exclusion of the remainder of the table, one obtains *conditional* information on that one row or column. Students should learn to dissect tables according to the joint, marginal, and conditional information that the table contains.

A key idea that permeates all of statistics is the notion of *association* between variables. Possible associations between two categorical variables (like gender and political party affiliation) can often be spotted when the frequencies are recorded in a two-way table. Possible associations between a categorical and a measurement variable (like gender and height) can be seen by parallel box plots. Associations between two continuous measurement variables (like height and weight) can best be seen in a scatterplot. This idea of looking for possible associations is not natural for students. They must be encouraged to find many examples of these types of comparisons and be provided with opportunities to discuss, both orally and in writing, the kinds of associations they see. Again, professional noticing comes into play.

PLANNING A STUDY

Answers to specific questions of a quantitative nature do not come about easily! Students should have some idea of how difficult it is to answer questions like "What percentage of voters think the school year should be increased in length?" or "Does a high-cholesterol diet increase the chance of a heart attack?" Even though they may never conduct surveys or experiments of their own, they should be aware of the key features of a well-designed study in order to have some basis for making decisions on the many issues that revolve around such studies. This understanding can begin in the upper elementary grades and can be deepened and broadened throughout the middle- and high-school years.

What Constitutes a Good Sample Survey?

As was observed in the introduction, surveys play a critical role in determining information that guides many aspects of our lives. From health care options to food options to entertainment options, much that is available to us was determined by a survey. Perhaps the best way to see the key elements of a good survey is to look at a hypothetical example.

School administrators want to know how many students will want parking spaces for automobiles next year. How can they get reliable information on this question? One way is to ask all of the returning students, but even this procedure would be somewhat inaccurate (why?) and will be very time-consuming. They could take the number of spaces in use this year and assume next year's needs will be about the same, but this method will have problems as well. A simple technique that works very well in many cases is to select a sample from those students who will be attending the school next year and ask each of them if they will be requesting a parking space. From the proportion of "yes" answers, an estimate of the number of spaces required can be obtained.

The scenario just outlined has all of the elements of a typical sample survey problem. There is a question of "How many?" or "How much?" to be determined for a specific group of objects called a target population, and an approximate answer is to be derived from a sample of data extracted from the population of interest. Randomization plays a vital role in the selection of samples that produce good approximations. It is clear that the administrator would not want to sample only seniors on the parking issue. It is less clear, but still true, that virtually any sampling scheme that depends upon subjective judgements as to who should be included will suffer from sampling bias.

Once the administrators know who is to be in the sample, they still need to get the pertinent information from them. The method of measurement, that is, the questions or measuring devices used to obtain the data, should be free of measurement bias. The administrators could choose from a number of possible questions to ask students about their parking needs for next year:

Do you plan to drive to school next year?

Do you plan to drive to school on more than half of the school days next year?

Do you have regular access to a car for travel to school next year?

Will you drive to school next year even if school bus service is improved?

Each of these questions will result in different information. For example, the third question likely will result in an overestimate of the number of spaces needed.

Because it is so difficult to get good information in a survey, every survey should be pretested on a small group of subjects similar to those who will be in the final sample. The pretest not only helps improve the questionnaire or measurement procedures, but also helps determine a good plan for data collection and data management. For example, for which of the parking questions can the responses be separated into a mutually exclusive and exhaustive set of categories so that they can be coded easily for the data analysis phase? The data analysis should lead to clearly stated conclusions that relate to the original purpose of the study. The goal of the parking study is to estimate the number of spaces needed, not the types of cars students drive or the fact that auto theft may be a problem.

Key elements of any sample survey include the following:

1. State the objectives clearly.
2. Define the target population carefully.
3. Design the sample selection plan using randomization, so as to reduce sampling bias.
4. Decide on a method of measurement that will minimize measurement bias.
5. Use a pretest to try out the plan.
6. Organize the data collection and data management.
7. Plan for careful and thorough data analysis.
8. Write conclusions in light of the original objectives.

Armed with an understanding of these key elements, students can properly critique a survey conducted by others or begin to design a good survey of their own.

Opinion polls are sometimes used to manipulate more than to inform, so it is essential that people know how to judge the merits of a poll. When they read a survey such as the one about the Holocaust, people should keep in mind three questions:

What was asked?

Who was asked?

How were they asked?

What Constitutes a Good Experiment?

The results of surveys are seen in practically every daily newspaper or weekly magazine. Less ubiquitous but perhaps more important are experiments, which are conducted to determine the effectiveness of "treatments" or to determine a causal relationship between variables. Most students will not conduct experiments of their own after leaving high school. However, many aspects of their lives will be influenced by experiments conducted by others. When a physician says "I prescribe medication A rather than medication B," an alert patient ought to ask penetrating questions as to why. When a local government decides to replace the current recycling system with a more efficient one, a concerned citizen ought to again ask intelligent questions as to why. The background information on choices such as these generally comes from experiments—good or bad.

Parents and teachers admonished students to turn off the radio while doing homework. Does listening to music while doing homework help or hinder? Only a carefully planned experiment can answer this question. Suppose a teacher plans such an experiment. She will have some students study with the radio on and some study with the radio off. However, the time of day that the studying takes place could affect whether or not the radio helps or hinders studying. So, she will have some students study in the afternoon with the radio on and some study in the afternoon with the radio off. Other students will study in the evening, some with the radio on and some with the radio off. The measurements on which the issue will be decided (for now) will be the scores on a quiz.

Because a radio might produce different results for boys and girls, perhaps the teacher should control for gender by making sure that both boys and girls are selected for each of the four treatment combinations. The native ability of the students might have some affect on the outcome as well. All of the students are from an honors history course, so it would be difficult to differentiate on ability. Therefore, the teacher will randomly assign students to the four treatment groups in the hope that any undetected differences in ability, and other uncontrolled factors, will balance out in the long run.

The teacher's study has most of the key elements of a designed experiment. The goal of an experiment is to measure the effect of one or more treatments on experimental units appropriate to the question at hand. Here, there are two main treatments, the radio and the time of day that study occurs. Another variable of interest is the gender of the student (the experimental unit in this case). This variable is directly controlled in the design by making sure we have data from both genders for all four treatment combinations. The variable "ability" cannot be controlled as easily, so the assignment of students to treatments is randomized to reduce the possible biasing effect of ability (and other background variables) on the response comparisons.

Key elements of any experiment include the following:

1. Clearly define the question to be investigated.
2. Identify the treatments.
3. Identify other important variables that can be controlled.
4. Identify important background (lurking) variables that cannot be controlled but should be balanced by randomization.
5. Randomly assign treatments to the experimental units.
6. Decide on a method of measurement that will minimize measurement bias.
7. Organize the data collection and data management.
8. Plan for careful and thorough data analysis.
9. Write conclusions in light of the original question.
10. Plan a follow-up study to answer the question more completely or to answer the next logical question on the issue at hand.

Similar to the case of sample surveys, the questions for people evaluating a study to keep in mind are:

What were the treatments?

Who (or what) was treated?

How were the treatments applied?

How were the results measured?

Key Concepts of Planning a Study

By the time students graduate from high school, they should be familiar with the following concepts of planning a study. (There are many possible orders of presentation for these topics.)

   I. Methods of data collection

      Census

      Sample survey (probability sampling)

Experiment (randomized, controlled)

Observational study (nonrandomized study)

Using archived data

II. Planning, conducting, and critiquing surveys

Role of randomization in sample selection; sampling bias

Questionnaire design; measurement bias

Sources of variability (sampling, measurement)

Reliability and validity

III. Planning, conducting, and critiquing experiments

Role of randomization in assigning treatments; bias and confounding

Comparative experiments; treatment versus control

Interaction

IV. Planning, conducting, and critiquing observational studies

Nonrandom assignment of "treatments"

Stratification to reduce confounding

DATA ANALYSIS

Once a study is designed around a specified question or problem and data are collected, the analysis of the data can begin. Data analysis is the process of

- Plotting the data to study the shape of the data distribution
- Summarizing key features of the data such as center and variability
- Looking for unusual patterns or unusual data points
- Looking for possible associations between variables
- Deciding which, if any, inference procedures are appropriate and carrying them out
- Making conclusions in light of the original question or problem
- Summarizing what further work should be done in a follow-up study

The following examples illustrate the main points of any data analysis problem.

What are average teacher salaries for the states of the United States and do they vary from region to region? This question can be answered by looking up the average teacher salaries as provided by the Statistical Abstract of the United States for 1992. Figure 1.1 provides a stem-and-leaf plot of these salaries (in thousands of dollars), which shows the skewed shape of the distribution. The boxplot of average salaries by region (1 = Northeast, 2 = Midwest, 3 = South, 4 = West) show that they do, indeed, vary by region.

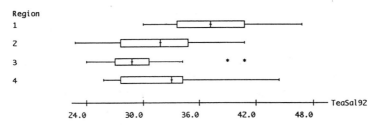

```
Stem-and-leaf of TeaSal92  N = 51
Leaf Unit = 1.0
      2 |3
      2 |445
      2 |666677777
      2 |8889999
      3 |0000111
      3 |333333
      3 |444445
      3 |667
      3 |89
      4 |0111
      4 |3
      4 |4
      4 |7
```

2|3 = $23,000

Boxplot of TeaSal92 by Region

Numerical Summaries of TeaSal by Region

	Region	N	MEAN	MEDIAN	STDEV
TeaSal92	1	9	37.80	37.30	5.33
	2	12	31.53	32.00	5.20
	3	17	30.09	29.00	4.61
	4	13	32.62	33.10	5.48

FIG. 1.1. Average teacher salaries, 1992.

This was to be expected; we would not expect the salaries in the four regions to be exactly the same. Inferential procedures can help answer the question: "Can these differences reasonably be attributed to chance or should we look for some other explanation?" Suggestions for follow-up studies might involve comparing salaries with other variables such as dropout rates and cost-of-living indices or obtaining current data to see if average salaries have increased.

Does having a job, in addition to being a full-time student, affect the number of hours a student studies per week? A sample of undergraduate students with jobs and a sample of students without jobs produced the boxplots of Fig. 1.2 for the variable "number of hours of study in a typical week." It appears that the students with jobs (coded 1) study less than the students without jobs (coded 2). In addition, there are some unusually large study hours in each group. The data set here consists of two rather small random samples from the large undergraduate student body at a major university. Thus, standard inferential procedures can be used and, in this case, they show that no statistically significant difference between the means

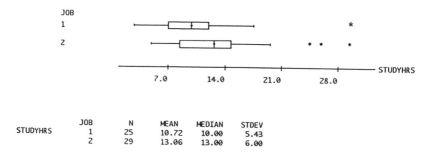

STUDYHRS	JOB	N	MEAN	MEDIAN	STDEV
	1	25	10.72	10.00	5.43
	2	29	13.06	13.00	6.00

FIG. 1.2. Weekly study hours of undergraduates.

has been established. A difference this large could quite likely have come from a population in which there is no difference between the average weekly study hours for the two groups.

By the time they graduate from high school, students should realize that data analysis often reveals patterns that look nonrandom. In these cases, inferential procedures should be used to decide whether or not the pattern reasonably can be attributed to chance. If so, then one should not make too much of the pattern. If not, then it is reasonable to hypothesize explanations other than chance and back them up with additional facts or, if needed, design a survey or experiment to try to find an explanation.

Key Concepts of Data Analysis

By the time students graduate from high school, they should be familiar with the following concepts and techniques of data analysis. (Again, there are many possible orders of presentation.)

I. Exploring univariate data: looking for patterns in measurement data
    Graphical: dot plot, number-line plot, stem plot, histogram
    Tabular: spreadsheet
    Center and spread
    Clusters and gaps
    Outliers and other unusual features
    Shape: symmetry versus skewness
    Effect of changing scale
II. Exploring univariate data: summarizing key features of measurement data
    Box plot; IQR and definition of outlier
    Measuring center and spread:
        Median and interquartile range for any distribution
        Mean and standard deviation for symmetric distributions

III. Comparing data sets; looking for association (measurement vs. category)

　　Graphical: dot plots, back-to-back stem plots, parallel box plots

　　Tabular: spreadsheet

　　Comparing centers and spreads

　　Within-group and between-group variation

IV. Exploring bivariate data; looking for association (measurement vs. measurement)

　　Scatterplots

　　Fitted line as a description of trend

　　Residuals, residual plots, and influential observations

　　Correlation

V. Exploring two-way frequency tables; looking for association (category vs. category)

　　Marginal and joint relative frequencies

　　Conditional relative frequencies

VI. Exploring time-series data; looking for trends

RELATIONSHIP OF PROBABILITY TO STATISTICS

Probability and inferential statistics are two sides of the same coin. By observing a random sample from a population, one can use inferential statistics to predict characteristics of that population. A typical question in inferential statistics is:

> In a random sample of 1,500 voters, 62% approve of the job the president is doing. What is the percentage of all voters who approve of the job the president is doing?

By observing a population, we can use probability to predict characteristics of random samples. A typical question in probability is:

> If a random sample of 1,500 voters is drawn from a population in which 60% approve of the job the president is doing, what is the probability that 62% of the sample will approve of the job the president is doing?

The second question must be answered before the first can be. Inferential statistics is based on probability.

The importance of probability in the secondary curriculum depends on more than its relationship to inferential statistics. Many of the decisions we make in modern life are probabilistic in nature. Probability is needed to understand lotteries, insurance, diagnostic medical testing, industrial quality control, weather forecasting, sports handicapping, genetics, and modern physics.

The Traditional Curriculum in Probability:
Probability as Combinatorics

The traditional "probability" strand in high school has been based on combinatorics. For example, the statistics and probability strand for high school students in the generally enlightened Mathematics Framework for California Public Schools Kindergarten Through Grade Twelve (1985) contains eight topics. The one listed first, and therefore the one that got the most attention, is "Use counting procedures to solve combinatorial problems." The framework also recommends an advanced one-semester elective course titled "Probability and Statistics." There are 11 suggested topics for such a course. The first is "Permutations" and the second is "Combinations." The California framework is not unusual; the probability chapter in many textbooks emphasizes these topics. Some units on probability consist almost entirely of contrived problems like this one:

> A committee of 4 is chosen at random from 10 men and 10 women. What is the probability that there are an equal number of men and women on the committee?

Note that the only probability used is the basic definition that when outcomes are equally likely, the probability of an event is the number of favorable outcomes divided by the total number of outcomes. Counting problems have been disguised as probability problems for so long that many people do not understand the difference.

It is true that combinatorial ideas are used in the development of some statistics and probability formulas, but so are many other mathematical topics. The slope–intercept form of a linear equation is fundamental in regression, but it is not listed as a statistics and probability topic.

Why has probability been redefined as counting? Mathematics teachers were trained in mathematics, not in statistics, and the counting formulas come more naturally than probability topics with their intimate relationship to statistics. Teaching permutations and combinations is more mathematical than supervising students setting up a computer simulation or discussing the law of large numbers, sampling in election polls, or what it means when the weather forecaster says there is a 40% chance of rain.

In addition, many teachers believe that students who will take a statistics course in college need to know the counting approach to probability. However, an examination of introductory college textbooks shows that very little combinatorial probability is necessary for the study of statistics. Questions like the one just given about the committee occasionally appear in a college discrete math course; they rarely appear in a statistics course.

Counting is beautiful and useful mathematics, but it is not probability. The NCTM Curriculum and Evaluation Standards for School Mathematics (1989) lists "solve enumeration and finite probability problems" in the discrete mathematics standard of the Grades 9–12 curriculum, not in the probability standard. The probability standard recommends: "At this level, the focus of instructional time should be shifted from the selection of the correct counting technique to analysis of the problem situation and design of an appropriate simulation procedure."

The Quantitative Literacy Project and others have had quite a bit of success in introducing simulation into the probability curriculum (Burrill, 1993; Scheaffer, 1993). But, in general, the simulations have been used to solve counting problems. The reform movement must build simulation into deeper studies of the distributions of random variables.

Teaching Probability, Not Combinatorics

A curriculum that emphasizes probability rather than combinatorics should have the following components.

1. *Probability should be presented as the study of random events.* Probability is often defined as the study of randomness, suggesting that students should have a wide range of experience with random events. Students should understand the law of large numbers and see how the percentage of successes can stabilize over many trials whereas the number of successes fluctuates more and more wildly about the expected number of successes. They should have watched sampling distributions grow as random samples are taken and then observed the approximate normality of the sampling distribution. Studying randomness, rather than probability formulas, will help students feel less mystified by real-world phenomena.

2. *The unifying thread throughout the probability curriculum should be the idea of distribution.* In the traditional probability curriculum, students are given no strategy other than "select the correct formula." No wonder they find the subject difficult and easily fall victim to "paradoxes." (Many teachers and their more mathematically inclined students love to grapple with these paradoxes—witness the recent fun over "Ask Marilyn.")

The unifying thread that underlies all of probability is the idea of distribution. For example, if a coin is tossed 100 times and the number of heads is counted, the probability distribution of the number of heads looks like the graph in Fig. 1.3. Probability questions should require students to observe the entire distribution rather than just the height of one bar. Instead of asking, "If you flip a coin 100 times, what is the probability of getting exactly 65 heads?" we could ask, "You are given a coin by a magician. You flip it 100 times and get 65 heads. Would you suspect the coin wasn't fair?"

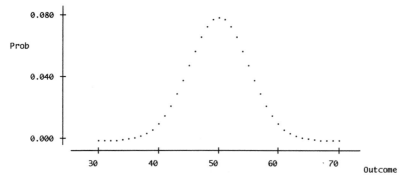

FIG. 1.3. Binomial distribution with $n = 100$, $p = 5$.

3. *Probability distributions typically should be constructed by simulation.* The college course in probability invariably is built around the standard probability distributions, which are constructed theoretically. Because high-school students don't have the mathematical tools to construct them, probability distributions have been absent from the high-school curriculum, with the occasional exception of the binomial distribution. But high-school students can construct these distributions by simulation.

Students learn more from a simulation than they do from using a formula to compute an answer. They see a probability distribution develop. Simulations are a natural way to begin to learn mathematical modeling. But most importantly, simulation builds students' confidence in their ability to handle any probability problem that might come up.

4. *Students' intuition about probabilistic events should be developed so that they can estimate probabilities and assess the reasonableness of results.* The only way students will be able to develop their intuition about probability is if they have experience with the basic probability distributions. Students who have studied probability distributions have a picture in their heads when asked questions such as, "If you flip a coin 100 times, what is the probability of getting 65 or more heads?" They can estimate the answer ("very small"). They know it is the same as the probability of getting 35 or fewer heads. They can visualize how the distribution changes if the probability of heads is 0.45 rather than 0.50.

The following typical probability problem is not hard to solve theoretically, although there are many ways to go wrong:

> According to the U.S. Postal Service, 96% of first class mail is delivered on time. Assuming that the letters are independent of one another, what is the chance that at least one of five letters is delivered late?

Many of today's students could use a formula to find the answer, but few have any intuitive grasp of this situation. Students with experience constructing probability distributions using simulation will understand the variability from sample to sample and be able to visualize the skewed binomial shape of the probability distribution: Usually, all five letters will arrive on time; four letters would be the next most likely; an occasional extremely unlucky person will have all five delivered late.

Students should realize the consequence of giving a somewhat unlikely event many chances of happening. For example, because millions of people mail five letters, eventually someone will have all of their letters delivered late and the postal service will get some bad publicity.

5. *Every student should learn the language and basic formulas of probability.* Students should know the language of probability (conditional probability, independence, equally likely, mutually exclusive). They should know how to calculate simple expected values and infer how insurance companies and lotteries make money. Students should understand the multiplication and addition rules.

6. *Misconceptions about probability should be confronted head-on.* Researchers have documented that people have many deep-seated beliefs about probability that are wrong. Only 56% of 11th graders know that if a fair coin lands tails up four times in a row, then it is equally likely to land tails up or heads up on the next toss (Brown, Carpenter, Kouba, Lindquist, Silver, & Suafford, 1988). People underestimate the length of the longest run of successes in a sequence of binomial trials. Consequently, they feel the need to find a special cause for these runs ("streaks" and "slumps"), especially in sports. Misconceptions tend to persist with age and in spite of formal training in probability. Researchers recommend that teachers should be forthcoming about the well-known misconceptions—tell students about them, demonstrate that they are false by confronting students with simulations, and remind students that they must always be suspicious of their first impulse when confronted with random events. The works by Falk and Konold (1992), Konold (1989), Kahneman, Slovic, and Tversky (1985), and Shaughnessy (1992) expand on both the importance and the difficulty of probabilistic reasoning.

Key Concepts of Secondary-School Probability

Probability as a relative frequency can be introduced in the elementary grades (especially with games of chance) and then expanded to the notions of probability distribution and expected value as the maturity of the students increases. By the time students graduate from high school, they should be familiar with the following concepts of probability.

I.   Analyzing a probability distribution given in table form
Basic vocabulary (mutually exclusive, independent, random variable)

Expected value and standard deviation from a one-way table

Histogram of the probability distribution from a one-way table

Conditional probability from a two-way table

II. Constructing discrete probability distributions by simulation

Use of random digit tables and the computer

Legitimacy of simulation; real-life uses of simulation

Estimate of error when using simulation/sample size needed

Important discrete probability distributions (binomial, geometric, discrete uniform, longest run of successes)

Shape of probability distributions (description, relationship to parameters)

III. Constructing discrete probability distributions by mathematical theory

Sample spaces for simple experiments (rolling two dice, flipping three coins)

Addition rule

Multiplication rule, tree diagrams

Binomial and geometric distributions

IV. Continuous probability distributions

Use of area to represent probability

Important continuous probability density functions (rectangular, normal)

Mean (expected value) (graphical interpretation as balance point)

Standard deviation (graphical interpretation)

V. Detecting and simulating random behavior

Heuristic tests for randomness (the digits of pi, hot hand, draft lottery)

Use of random digit generators in calculators and computers

VI. Law of large numbers

Notion of convergence of sample paths

Importance of independence

VII. Sampling distributions (covered in detail in the inferential reasoning outline)

Sampling distribution of the mean

Sampling distribution of other statistics (possibly non-normal)

VIII. Common misconceptions about probability

The law of averages (believing that, e.g., in coin tossing the number of heads and the number of tails must eventually even out)

Underestimating the length of runs in binomial trials

Underestimating the probability of a rare event that is given many opportunities to occur

Understanding the relationship between variability and sample size

Underestimating probability of a match (birthday paradox)

IX. Applications

Insurance

Lotteries and games of chance (including random walks, gambler's ruin)

Risk assessment

Queues

Stochastic versus deterministic models in physics

Diagnostic testing (screening for a disease such as AIDS)

Weather forecasting

Sports handicapping

Genetics

INFERENTIAL REASONING

Probability reasons from the general to the particular (from the population to the sample), whereas statistical inference reasons from the particular to the general (from the sample to the population). Inferential reasoning is alien to most students, so the topic must receive concentrated attention in the school curriculum, first informally and then more formally. The basic building block is the notion of probability distributions, as outlined in the preceding section. The following illustrates the type of argument that a high-school graduate should be able to construct.

"If Women Ran America" is the provocative headline of an Associated Press article on a poll comparing women's and men's attitudes on various public issues. In a poll commissioned by *Life* magazine, the Gallup Organization randomly sampled 614 women and 608 men from across the United States. The results appear to show substantial differences between men and women, especially on issues surrounding the workplace, crime, and social behavior. Three of the statements used in the poll and the sample percentages of each gender who say they agree with each statement are provided in Table 1.2.

The objective of polls such as the one just described is to gauge the opinions of the nation, not the opinions of 614 women and 608 men who happened to be in the sample. The problem, then, is to decide what population percentages could have given rise to these sample percentages. Could the population percentage of women who agree with Statement II be as low as 50% or even

TABLE 1.2
Sample From an Attitude Poll: Percent Agreeing With Statement

	Female	Male %
I. Discrimination in promotion is a serious problem.	50	33
II. Government should make fighting crime and violence an extremely important priority.	55	46
III. The justice system is not tough enough on drunken drivers.	76	58

45%? Is it, in fact, true that more women than men agree with all three statements? Statistical inference helps answer these questions.

Simulation Approach

One part of the original problem shows 55% of a random sample of 614 women agreeing with Statement II on crime and violence. What is a reasonable set of possible values for the true proportion of women in the U.S. population who agree with this statement? More specifically, is 50% a reasonable guess for a population percentage, given these sample results? An approximate answer can be found through simulation, making use of the binomial distribution that was introduced in the section on probability. A way, then, to construct the simulation is to generate 100 values from the binomial (614, .5) distribution, and then divide each of these outcomes by 614 to turn them into binomial proportions. Many computer software packages and graphing calculators have a built-in program to generate outcomes from binomial distributions.

If it turns out the .5 is a reasonable choice for the population proportion, what about .4 or .6? Simulated sampling distributions would have to be constructed for a variety of possible choices before a reasonable set of possible values for the true population proportion could be selected. Simulated sampling distributions, each containing 100 values of the sample proportion, are shown in Fig. 1.4. Students now can notice that a choice of .45 for the population proportion is not reasonable, since .55 seems to have little chance of occurring in the samples of size 614 from this population. A choice of .50 might seem reasonable, but only one sample selected from a population with 50% had a sample proportion of .55 or more. On the upper end, the sampling distribution for .57 shows that a sample proportion of .55 is quite likely to occur in this situation. A choice of .60 seems to be just a little too large, since the approximate sampling distribution just touches .55. So, a reasonable set of possible values for the true population proportion is (.51, .59), an approximate *confidence interval.* After much experience with simulations of this type, in varying contexts, students begin to develop

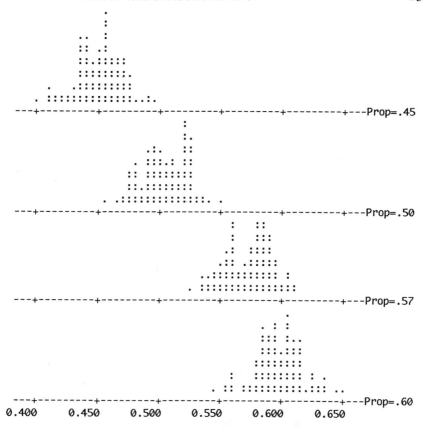

FIG. 1.4. Sample proportions, binomial with $n = 614$.

some "feel" for reasonable values of population parameters, and a more formal treatment of confidence intervals can begin.

Theoretical Approach

Careful study of confidence intervals constructed by simulation leads to two important observations:

1. The center of the interval is at the observed value of the sample proportion.
2. The length of the interval is inversely proportional to the square root of the sample size.

These two facts are true for confidence intervals on population proportions as long as the observed proportions are not too close to either 0 or 1 and the sample size is reasonable large. Studying the behavior of confidence intervals for proportions for widely ranging sample sizes and values of the true population proportion leads to the conclusion that these intervals can be closely approximated by the following formula:

$$p \pm 2\sqrt{\frac{p(1 - p)}{n}}$$

where p denotes the observed sample proportion. For confidence intervals written in the form $p \pm B$, the B is called the *margin of error*. The interpretation of this confidence interval is that out of every 100 such confidence intervals constructed, 95 of them are expected to contain the true population proportion. These formulas can be developed from theoretical considerations in advanced mathematics courses, but should be developed only as approximations to the simulations for most students. Students should observe how the length of the interval changes with the sample size and with observed values of p.

After study of confidence intervals for single parameters (proportions and means being the two most common), comparisons between parameters can be made. This leads quite naturally to tests of significance, as in answering the question, "Is it true that more females than males agree with Statement II?"

Key Concepts of Statistical Inference

The notions of statistical inference are subtle and cannot be fully developed until high school. By the time a student leaves high school, however, he or she should be familiar with the following ideas.

I. Confidence intervals for proportions and means
        Construction and verification by simulation
        Sampling error, "margin of error"
II. Tests of hypotheses
        Addressed through confidence intervals
        Addressed through simulation (informal notion of a p value)
III. Modeling bivariate data
        Least squares line
        Heuristic inference for the slope
        Heuristic inference for the correlation coefficient

CONCLUSION

The mathematics curriculum of the schools should emphasize number sense, study planning, data analysis, probability from a distributional perspective, and statistical inference introduced from an intuitive perspective. The key concepts for these topics are summarized in the lists at the end of the preceding four sections. Moore (1990) offered a slightly different perspective on these topics. Several organizations recently offered guidelines that include the teaching of statistics and probability. The Appendix contains excerpts from the statements of the American Statistical Association (1991), the National Council of Teachers of Mathematics (1989), the 1992 Mathematics Framework for California Public Schools, and the American Association for the Advancement of Science (1993). There clearly are differences in emphases and expectations among these policy statements and the key concepts presented in this chapter. It is also clear, however, that there is an overall agreement in philosophical and pedagogical approach, and in the notion that effectively teaching statistics to all students is important and possible.

Many of the basic ideas discussed in this chapter can be taught beginning in the elementary grades. As these ideas are revisited and embellished in the later grades through increasingly sophisticated problems appropriate for students' ages and interests, students have opportunities to reinforce and clarify their understanding of the basic ideas while adding complementary ones. Statistics involves subtle reasoning, and revisiting and reinforcing the basic concepts are vital for students to develop profound reasoning skills by the end of Grade 12.

Statistics should not be taught in isolation from the remainder of the mathematics curriculum or from the curricula in science and social science. Statistics provides rich motivation and illustration for much of traditional mathematics. It also provides avenues for connecting experimental science and quantitative aspects of social science to mathematics. Statistics within mathematics is an ideal way to improve a student's knowledge of both and thereby to bring the student's quantitative reasoning skills to the level that modern society demands.

Statistics in the school mathematics curriculum is not a new idea. Throughout this century, many have proposed that statistical reasoning should play a more prominent role in the school curriculum in general and in the mathematics curriculum in particular. The first NCTM Yearbook, published in 1926, included this passage:

> One of the most significant evidences that the importance of mathematics is permeating the whole fabric of modern life is shown in the recent unparalleled development of the use of statistical methods in the study of quantitative

relations in almost every department of investigation. This appears in the simplest form in all the proposed new curricula for the junior high schools. (Slaught, 1926/1995, p. 192)

The effort to improve statistics education in the schools can succeed now, where it has failed in the past, because of an ever-increasing use of data in the modern world, an abiding interest in making mathematics more practical, and the advent of data-analytic techniques and attendant computer software that make statistical analyses fun and easy to conduct. All the ingredients are available to allow the ideas proposed above to come to fruition, and relatively quickly, if effort is concentrated in that direction.

REFERENCES

American Association for the Advancement of Science. (1993). *Benchmarks for science literacy.* New York: Oxford University Press.

American Statistical Association. (1991). *Guidelines for the teaching of statistics.* Alexandria, VA: Author.

Brown, C. A., Carpenter, T. P., Kouba, V. L., Lindquist, M. M., Silver, E. A., & Swafford, J. O. (1988). Secondary school results for the Fourth NAEP Mathematics Assessment: Discrete mathematics, data organization and interpretation, measurement, number and operations. *Mathematics Teacher, 81,* 241–248.

Burrill, G. (1991, August). Quantitative literacy—Implementation through teacher inservice. In *Proceedings of the Third International Conference of Teaching Statistics* (pp. 50–55). Voorburg, The Netherlands: International Statistical Institute Publications.

Crossen, C. (1994). *Tainted truth: The manipulation of fact in America.* New York: Simon & Schuster.

Falk, R., & Konold, C. (1992). The psychology of learning probability. *Statistics for the 21st Century, MAA Notes, 26,* 151–164.

Kahneman, D., Slovic, P., & Tversky, A. (1985). *Judgment under uncertainty: Heuristics and biases.* Cambridge: Cambridge University Press.

Konold, C. (1989). Informal conceptions of probability. *Cognition and Instruction, 6,* 59–98.

Mathematics frameworks for California public schools, kindergarten through grade twelve. (1992). Sacramento: California Department of Education.

Moore, D. S. (1990). Uncertainty. In L. A. Steen (Ed.), *On the shoulders of giants* (pp. 95–137). Washington, DC: National Academy Press.

National Council of Teachers of Mathematics. (1989). *Curriculum and evaluation standards for school mathematics.* Reston, VA: Author.

Scheaffer, R. (1991). The ASA-NCTM Quantitative Literacy Program: An overview. In *Proceedings of the Third International Conference of Teaching Statistics* (pp. 45–49). Voorburg, The Netherlands: International Statistical Institute Publications.

Shaughnessy, J. M. (1992). Research in probability and statistics: Reflections and directions. In *Handbook of research on mathematics teaching and learning* (pp. 465–494). Reston, VA: National Council of Teachers of Mathematics.

Slaught, H. E. (1995). Mathematics and the public. In *A general survey of progress in the last twenty-five years* (pp. 186–193). Reston, VA: National Council of Teachers of Mathematics. (Original work published 1926)

APPENDIX: RECOMMENDATIONS FROM OTHERS
ON TEACHING PROBABILITY

The *Guidelines for the Teaching of Statistics* from the American Statistical Association (ASA; 1991) even in Grades 9–12 did not go much beyond what we would expect from a reasonably robust middle school program:

> Students should be able to identify the probability of an event as a number between 0 and 1, express this as a fraction, decimal or percent, whichever is appropriate, and recognize the relation between area and probability. Terms such as likely, equally likely, and unlikely should be used to describe everyday events as well as the outcomes of an experiment. Students should be able to distinguish between independent and dependent events and recognize and assign probabilities to complementary events and to two or more related events. For simple experimental situations involving counts, students should be able to compute expected values. They should also be able to compare the theoretical probability and an estimated empirical probability for the same event.
>
> Students should investigate probabilities empirically by experimentation with devices that generate random outcomes, estimate probabilities through collecting and observing real data, and use probability to interpret tables and graphs. They should understand the concept of simulation and use it as a technique to solve real problems. Students should recognize the variability in estimating a probability and how that variability decreases as the number of simulations increases. The concept of sampling distributions should be investigated experimentally for a variety of situations and lead to an introduction to the normal distribution. (p. 38)

The NCTM *Curriculum and Evaluation Standards for School Mathematics* (1989) went further than the ASA:

> In grades 9–12, the mathematics curriculum should include the continued study of probability so that all students can
>
> > use experimental or theoretical probability, as appropriate, to represent and solve problems involving uncertainty;
> >
> > use simulations to estimate probabilities;
> >
> > understand the concept of a random variable;
> >
> > create and interpret discrete probability distributions;
> >
> > describe, in general terms, the normal curve and use its properties to answer questions about sets of data that are assumed to be normally distributed;
>
> and so that, in addition, college-intending students can
>
> > apply the concept of a random variable to generate and interpret probability distributions including binomial, uniform, normal, and chi square. (p. 171)

In addition, the discrete math strand of the Standards recommends that all students can "solve enumeration and finite probability problems."

The 1992 *Mathematics Framework for California Public Schools* was the most ambitious, but more combinatorial:

> To make sense of experiments involving chance, students employ basic ideas of probability: randomness, independent events, mutually exclusive events, complementary events, equally likely events, and expected value. They can find the probability of an event, using ideas and methods that include relative frequency, simulation, sample space, tree diagrams, tables, and geometric models; and they can use the basic operations on probabilities, including conditional probabilities. By experiment and simulation the binomial distribution is constructed, and its application to real phenomena and random walks is developed. In addition, the difference between empirical and theoretical probabilities is clarified. Students employ counting principles and methods to solve problems that involve combinations, permutation, and Pascal's triangle.
>
> In the core curriculum probability is developed in the context of its uses in interesting situations: odds, games, insurance, the lotteries, genetics, and forecasting. . . .
>
> In the post-core curriculum, further development of statistical methods should include normal distribution and its relationship to the binomial, the central limit theorem. . . .
>
> Understanding of theoretical probability should be extended to include methods for determining expected value. (pp. 148–149)

Recommendations From Others on Teaching Statistics

The *Guidelines for the Teaching of Statistics* from the American Statistical Association (1991), following similar guidelines for the earlier grades, focused on the following outline of topics for Grades 9–12.

I. Exploring Data
   A. Understanding the problem
   B. Gathering and exploring data
   C. Organizing and representing data
   D. Describing data
   E. Interpreting data
II. Statistical Inference
   A. Planning a survey or experiment
   B. Inferential statistics
   C. Statistics in society (pp. 35–37)

All of the topics in part I are recommended for elementary and middle grades as well, with less sophisticated examples. Topics A and C of part II are also recommended for the elementary and middle grades, with some suggestions for an informal look at inference for those grades.

The NCTM *Curriculum and Evaluation Standards for School Mathematics* (1989) presented a more specific list of goals:

In grades 9–12, the mathematics curriculum should include the continued study of data analysis and statistics so that all students can

construct and draw inferences form charts, tables, and graphs that summarize data from real-world situations;

use curve fitting to predict from data;

understand and apply measures of central tendency, variability and correlation;

understand sampling and recognize its role in statistical claims;

design a statistical experiment to study a problem, conduct the experiment, and interpret and communicate the outcomes;

analyze the effects of data transformations on measures of central tendency and variability;

and so that, in addition, college intending students can

transform data to aid in data interpretation and prediction;

test hypotheses using appropriate statistics. (p. 167)

The 1992 *Mathematics Framework for California Public Schools* built on the NCTM Standards in developing more detailed objectives:

Statistics and probability build on collecting, analyzing, representing, and interpreting data from real and simulated situations. (pp. 148–149)

Many of the ideas and methods of statistics are based on the concept of frequency distributions. Students need a sense of how frequency distributions are created and interpreted in a variety of situations.

Using statistical methods, high school students should be able to design and carry out a simple but well-thought-out research project. Data can be collected by survey methods or experiment.

Techniques of data analysis should include measures of central tendency (mean, median, mode); dispersion (variance, standard deviation, range); and relationship (correlation and regression). An informal study of the use of inferential procedures, including hypothesis testing and confidence intervals, are also appropriate. (pp. 148–149)

The construction, analysis, and interpretation of graphs, charts, and tables should be emphasized in the core. Forms of software, including graphing applications, spreadsheets, databases, and statistical packages, are basic tools for work in statistics. Students need experience with their use.

> The post-core curriculum should include confidence intervals, statistical significance, the meaning and use of correlation, regression, outliers, z-scores, standard deviation and other summary statistics. (pp. 148–149)

Recommendations From the Science Community

The American Association for the Advancement of Science (AAAS) *Benchmarks for Science Literacy* (1993) presented a compendium of specific science literacy goals. For topics dealing with data, chance, and uncertainty—words that for many seem not to have the negative connotations of "statistics" and "probability"—this report stated that "the principal intent is to make [high-school graduates] informed consumers, not producers, of data. For example, students should know that people can be alert to possible bias in choosing samples that others take but may be unable to take adequate precautions against bias in designing a study of their own" (p. 226).

The following are specific selected goals pertaining to different grade levels.

By the end of the 5th grade, students should:

> Know that summary predictions are usually more accurate for large collections of events than for just a few. Even very unlikely events may occur fairly often in very large populations. (p. 228)

> Know that spreading data out on a number line helps to see what the extremes are, where they pile up, and where the gaps are. A summary of data includes where the middle is and how much spread is around it. (p. 228)

> Recognize when comparisons might not be fair because some conditions are not kept the same. (p. 299)

By the end of the 8th grade, students should:

> Know that how probability is estimated depends on what is known about the situation. Estimates can be based on data from similar conditions in the past or on the assumption that all the possibilities are known. (p. 229)

> Know that comparison of data from two groups should involve comparing both their middles and the spreads around them. (p. 229)

> Know that the larger a well-chosen sample is, the more accurately it is likely to represent the whole. But there are many ways of choosing a sample that can make it unrepresentative of the whole. (p. 229)

> Be able to find the mean and median of a set of data. (p. 291)

> Be able to estimate probabilities of outcomes in familiar situations, on the basis of history or the number of possible outcomes. (p. 291)

> Be skeptical of arguments based on very small samples of data, biased samples, or samples for which there was no control sample. (p. 291)

> Be aware that there may be more than one good way to interpret a given set of findings. (p. 299)

By the end of the 12th grade, students should:

> Know that when people estimate a statistic, they may also be able to say how far off the estimate might be. (p. 230)
>
> Know that the middle of a data distribution may be misleading—when the data are not distributed symmetrically, or when there are extreme high or low values, or when the distribution is not reasonably smooth. (p. 230)
>
> Know that considering whether two variables are correlated requires inspecting their distributions, such as in two-way tables or scatterplots. A believable correlation between two variables doesn't mean that either one causes the other; perhaps some other variable causes them both or the correlation might be attributable to chance alone. A true correlation means that differences in one variable imply differences in the other when all other things are equal. (p. 230)
>
> Know that the larger a well-chosen sample of a population is, the better it estimates population summary statistics. For a well-chosen sample, the size of the sample is much more important than the size of the population. To avoid intentional or unintentional bias, samples are usually selected by some random system. (p. 230)
>
> Be able to use computer spreadsheet, graphing, and database programs to assist in quantitative analysis. (p. 291)
>
> Be able to compare data for two groups by representing their averages and spreads graphically. (p. 291)
>
> Notice and criticize arguments based on the faulty, incomplete, or misleading use of numbers, such as instances when (1) average results are reported, but not the amount of variation around the average, ... (p. 300)
>
> Check graphs to see that they do not misrepresent results by using inappropriate scales or by failing to specify the axes clearly. (p. 300)
>
> Wonder how likely it is that some event of interest might have occurred just by chance. (p. 300)
>
> Suggest alternative ways of explaining data and criticize arguments in which data, explanations, or conclusions are represented as the only ones worth consideration, with no mention of other possibilities. . . . (p. 300)

As an overall statement of the philosophy behind the specific goals such as these, the report states that:

> The distinction between ends and means should be kept in mind in all of this. The ultimate aim is not to turn all students into competent statisticians but to have them understand enough statistics to be able to respond intelligently to claims based on statistics; without the kind of intense effort called for here, that understanding will be elusive. (p. 229)

CHAPTER TWO

Statistics and Probability for the Middle Grades: Examples From *Mathematics in Context*

Gail Burrill
Thomas A. Romberg
University of Wisconsin–Madison

Today we live in a data-driven society and are confronted with data in everything—from work reports to the daily newspaper. Through use of probability and statistics, we can learn to deal with these data in a responsible way and to solve problems based on current and past data. Understanding statistics and probability enables people to reason from and make conclusions based on data, judge the quality of other people's conclusions, recognize the degree of uncertainty in any endeavor, and quantify that uncertainty.

Although we assume that students will have encountered some work with data in the primary grades, it is in the middle grades that these understandings begin to be developed in depth. This chapter describes the features of this important branch of school mathematics and characterizes how these features have been operationalized in *Mathematics in Context* (MiC) (National Center for Research in Mathematical Sciences Education & Freudenthal Institute, in press), a middle-school curriculum project funded by the National Science Foundation. To illustrate this project we have included some examples that typify student work from a few activities in the MiC units. The chapter concludes with a brief discussion of some issues concerning the implementation of this program in schools.

GOALS OF STATISTICS AND PROBABILITY FOR SCHOOL MATHEMATICS

From a school mathematics perspective, *statistics and probability* should be seen as a branch of mathematics dealing with the collection, analysis, interpretation, and presentation of information derived from data. Because

33

ideas in this domain have not commonly been taught in American schools, we started to develop the statistics and probability strand in MiC with the vision of this domain for the K–12 mathematics curriculum as portrayed by the National Council of Teachers of Mathematics in its *Curriculum and Evaluation Standards* (NCTM, 1989). For the middle-school grades, we needed to identify the "big" ideas from this branch of mathematics, suggest a sequence for the development of these ideas, indicate the important pedagogical notions related to these ideas, specify necessary related mathematical concepts and procedures, and decide what should be included in the MiC units.

In the NCTM *Curriculum and Evaluation Standards* (1989), statistics and probability are included in the curriculum content of each of the three grade levels (K–4, 5–8, 9–12). After reviewing these standards we decided that the probability and statistics units for the middle grades should enable students to understand the role of statistics and probability in their own lives as well as within the context of mathematics. We envisioned students using data collected from a variety of sources (census, samples, experiments, simulations), using graphs and numerical statistics to describe and summarize the variation inherent in that data, making conjectures about possible answers to questions derived from reflections on these descriptive statistics, and using probability to measure how uncertain a possible conclusion might be. Because this process is iterative (reflection on the problem, the data, and the conclusions often leads to new problems, or new insights into the given problem), students during the 4 years of mathematics in the middle grades need to confront a number of rich and varied problem situations that will allow them to develop from the preformal intuitive notions that they bring to such problems, toward more formal quantitative approaches they can use to examine more complex problems. We decided, in addition, that the emphasis at this level must be on concept building and reasoning, not on using rules and formulas. The formal procedures of testing hypothesis, developing complex statistical models, and so forth are best left to high-school and college curricula. Nevertheless, the experiences students have in the middle-level grades should both acquaint them with the need for and prepare them for the development of those more formal procedures.

In summary, we believe that statistics and probability for the middle grades should bridge the gap between students' intuitive and informal ideas about making conjectures from collections of data and the formal ways users of statistics analyze data.

STATISTICS AND PROBABILITY
IN *MATHEMATICS IN CONTEXT*

Mathematics in Context: A Connected Curriculum for Grades 5–8 (MiC) is a comprehensive mathematics curriculum comprising 40 units (10 at each grade level). The development of the MiC units has been through a collaboration between research and development teams of both the National Center for Research in Mathematical Sciences Education (NCRMSE) at the University of Wisconsin–Madison and the Freudenthal Institute at the University of Utrecht, The Netherlands, and a group of American middle-school teachers who piloted the units.

The content of the Statistics and Probability strand of the MiC materials mirrors closely that described in chapter 1 by Scheaffer, Watkins, and Landwehr in this volume. The developmental process and theoretical notions about instruction that underlie the MiC units, however, provide one example of how that content can be taught in classrooms. In creating a set of instructional units for statistics and probability in the middle grades, we were particularly concerned that students develop an understanding of the role and interpretation of statistics in contemporary society. Note that this focus is in sharp contrast with the traditional focus on computation that has been dominant in the past. To encourage students' growth in this area we rely on real-world contexts. Several chapters in this volume (Derry et al., chapter 7; Schwarz et al., chapter 9; and Lajoie et al., chapter 8) provide other examples of contextualized real-world problem solving.

We are aware that the user of statistics is involved in a complex set of activities that goes far beyond the use of the specific techniques often taught in statistics courses. Our assumption was that students should begin their investigations with the context of a real-world problem that needs to be addressed. This initial step involves raising questions about the problem situation and attempting a hypothesis, but consensus was that hypothesis generation, a critical aspect of statistical reasoning, has rarely been taught. From this perspective, the initial instructional activity should be experientially real to students so that they engage in personally meaningful mathematical work (Gravemeijer, 1990). Students then must identify information needed to answer the questions. This process involves specifying variables, describing a model that reflects the relationships between variables, and either searching for existing data or collecting new data about the variables. These batches of data then need to be organized, and students address issues such as the differences between categorical data and numerical data. Only then do students, using a variety of descriptive and quantitative techniques, begin to make sense of the data. These techniques include of procedures for visualizing the data, calculating data summaries for central tendency and dispersion, producing tables and scatterplots for pairs of data, and so on. After creating such descriptions of data, students make decisions or predictions based on the data. At this point the importance of probabilistic reasoning becomes apparent. The role of probability in making rational decisions, however, is problematic unless sampling and design issues are addressed. Finally, students build a coherent case to answer questions based on the original information and analysis.

Throughout this sequence, MiC emphasizes that statistical concepts should connect to situations where they make sense in a "natural" way. Statistical and probability tools and concepts are not presented and learned as a long list of separate items, but rather as items connected to each other and to their applications in given situations. For example, measures of central tendency are connected to different graphic representations. The ability to think critically about a situation is emphasized throughout the entire development of the strand.

The Units

In MiC, the eight units in the statistics and probability strand are organized into two substrands: statistics and probability (Table 2.1). In most of these units, one central theme is used to organize a set of concepts. For example, the initial unit, Picturing Numbers (Boswinkel et al., 1997), does just that—it pictures numbers. Experiences in reading and making statistical pictures or graphs are closely connected to experiences students might have with statistics in newspapers and other media. In one exercise, students examine temperature data from a travel brochure for a set of resorts in the Caribbean (see Figs. 2.1 and 2.2). Students answer questions such as, "Based on the temperatures, where would you prefer to go for a vacation in January?" They then compare the plotted temperatures of three unnamed resorts. After figuring out which resorts match the three plotted temperature lines, students decide which of the three has, in their opinion, the best climate and explain their choices. In this instance, they are literally "picturing numbers" and using data to make decisions.

TABLE 2.1
MiC Statistics Units

Grade	Statistics	Probability
5	Picturing Numbers: simple graphic representations: informal numerical summaries of data	Take A Chance: fair decision making; probability introduced as chance
6	Dealing With Data: graphic representations, measures of center	
7	Statistics and the Environment: applications of graphic and numerical techniques	Ways to Go: formalization of probability based on frequencies
8	Insights Into Data: scatterplots, introduction to sampling and random samples	Great Expectations: expected value; relationship between samples and population
	Digging Numbers: statistics in context	

Average Temperatures (°F)

	J	F	M	A	M	J	J	A	S	O	N	D
Jamaica	86	87	87	88	90	90	90	90	89	89	89	87
U.S. Virgin Islands	80	81	82	83	88	88	90	90	88	87	86	86
St. Martin/St. Kitts	80	81	82	83	86	86	86	87	87	86	85	84
Aruba	83	84	84	86	88	88	88	41	91	90	89	86
Antigua	80	80	80	82	90	90	90	90	89	89	89	83
Grand Cayman	88	87	86	88	88	89	90	91	91	89	88	88
Nassau	76	76	78	80	84	88	89	90	88	84	81	79
Cancún	84	85	88	91	94	92	92	91	90	88	86	82
Cozumel	84	85	88	91	94	92	92	91	90	88	86	82
Los Cabos	73	74	79	83	88	93	95	93	92	89	82	74
Acapulco	88	88	88	88	89	90	91	91	90	90	90	89
Puerto Vallarta	76	77	81	85	83	88	93	95	93	92	89	75
Ixtapa	89	90	92	93	89	88	89	90	91	91	90	89
Mazatlán	73	74	79	83	84	92	94	92	92	90	85	71
Manzanillo	77	78	82	86	83	88	53	95	93	92	89	74

FIG. 2.1. Picturing temperatures from a chart. Reprinted by permission of the Encyclopaedia Britannica Educational Corporation.

Data Collection

Students are introduced to data and data collection activities in Take A Chance (Jonker, van Galen, et al., 1997) in grade 5. They toss coins and number cubes in the probability units and learn how to record and organize their results. Students conduct survey and compare their results to those in the text. Students also investigate bias in surveys and learn how results can be biased through questioning techniques, sampling strategies, and the measuring instruments themselves.

Data Visualization

Students begin to use graphic representations for data in Picturing Numbers (Boswinkel et al., 1997). They construct and interpret bar graphs, pictograms, and number-line plots. Students learn to label axes and choose appropriate scales. The connection between the graphs and the situations from which the data are collected is critical, and students are asked to think about a graph only as a way to understand the situation, not as an end in itself. Students add scatterplots, box plots, histograms, and stem-and-leaf plots to their collection of graphic tools.

These graphs are analyzed for advantages and disadvantages, and different displays are inspected to see what information can be obtained from them. Conclusions based on data and graphs are examined, and several activities focus on analyzing the way graphic representations are used in the media.

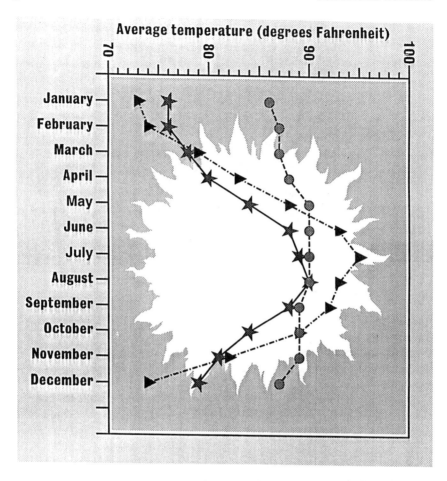

FIG. 2.2. Picturing temperatures from a graph. Reprinted by permission of
the Encyclopaedia Britannica Educational Corporation.

In the final unit, Digging Numbers (de Lange et al., in press) students are
given an archeological problem that allows them to choose among the
statistical techniques they have learned. The context involves the jaw bones
of approximately 70 ancient pike that archaeologists find in an old lake bed.
They want to know whether there is any similarity between ancient and
contemporary pike and try to estimate the length of the ancient pike. Data
from a nearby lake about the length and width measurements of contemporary
pike are supplied, and students are asked to use these data to resolve the
problems. Students can organize the information in a stem-and-leaf plot or a
number-line plot (Fig. 2.3). When the jawbone lengths are displayed in
side-by-side box plots or corresponding histograms, the answer to the first
question is quite clear: The ancient pike were smaller than contemporary pike.

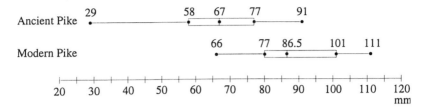

FIG. 2.3. Using a number-line plot to display jawbone lengths. Reprinted by permission of the Encyclopaedia Britannica Educational Corporation.

A scatterplot of the body and jawbone lengths of the contemporary pike reveals a linear trend that can summarized by a line (Fig. 2.4) and an equation, for example, $L = 6.3J + 63.2$. Students may use either the line or the equation to estimate the length of an ancient pike given the measure of the jawbone.

Numerical Characteristics

In Picturing Numbers (Boswinkel et al., 1997) in grade 5, students use numeric measures such as maximum, minimum, and mode. They are introduced to the mean as a way to describe the center of a data set. The mean is initially presented through a compensation strategy: To find the mean, students "redistribute" the cats owned by a number of families so that each family will have the same number of cats.

The mean is revisited in grade 6, and students learn to calculate the mean height for a very large sample of heights. The mean, mode, median, and quartile are related to graphic representations. The concept of variation is introduced informally, through range, quartiles, and the concept of the spread of a distribution. Students learn that the mean can be represented as the balance point of a distribution. They also learn that the mean alone is not an adequate way to think about data: They encounter two sets of temperatures, one for San Francisco and one for St. Louis, MO, that have approximately the same yearly mean (56°F), yet very different distributions (Fig. 2.5). Finally, students learn to think critically about the mean as a meaningful number for any given situation.

Reflection and Conclusions

Consistently throughout the units, students are asked to conjecture, make conclusions, and reflect on their work. They learn to use statistical evidence to support their arguments in sensible and meaningful ways and are asked to do so both orally and in writing. In grade 5, students draw conclusions from pictures and informally discuss how to do so reasonably. The focus in grade 6 is to draw conclusions about the relationships between the heights

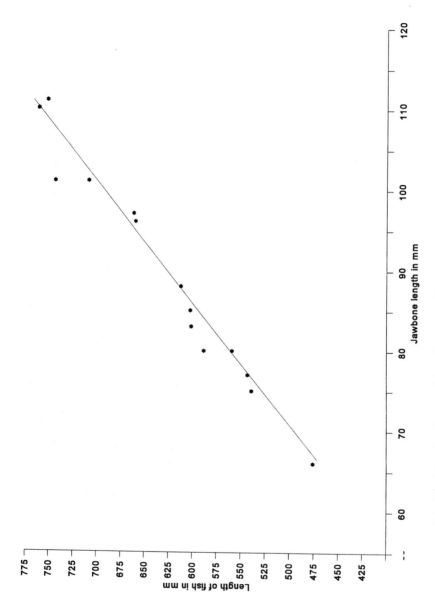

FIG. 2.4. Estimating length of the pike using a line graph. Reprinted by permission of the Encyclopaedia Britannica Educational Corporation.

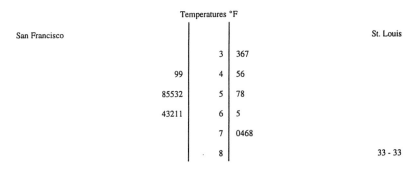

FIG. 2.5. Yearly means: investigating distribution. Reprinted by permission of the Encyclopaedia Britannica Educational Corporation.

of parents and those of their children and to explore how the use of statistical techniques can be used to justify conclusions. In grade 7, students reflect on population data and decide how to apply their conclusions to environmental planning on the island. In grade 8, students use their understanding of linear relationships to make predictions and are asked to reflect on the consequences of extrapolating beyond a given set of data.

The groundwork for formal statistical inference begins in grade 6, when concepts of sample and population are introduced, and the idea of representative samples is raised. In grade 7, students address the issue of how to obtain a random sample and how to use this concept in the data-gathering process. In one activity, students describe how a typical population uses electricity and then use this information to make predictions for the island. In grade 8, samples, populations, and the notion of bias in the sampling process are again revisited. Students also investigate questions about representative samples and sample size and their effect on the validity of conclusions. In order to determine whether *Harper's Index* is correct in stating that 80% of the prices in a supermarket end in a 9 or a 5, students use random numbers to simulate a sampling distribution and inspect the distribution for likely outcomes. They compare these results with those from an actual survey of prices in a local supermarket.

In the probability substrand the conceptual themes are descriptions of chance and measures of uncertainty.

Descriptions of Chance

In grade 5, students are informally introduced to the concepts of fair and unfair. They are asked to gauge fairness in simple situations involving spinners, dice, and coins. They explore a range for probability, moving from intuitive ideas of "sure" and "not sure" to the use of a probability ladder marked with fractions and percents. Students compare probabilities and approximate them using repeated trials. Simple counting strategies and tree diagrams are introduced to help students represent situations where they have to make two or three different choices. Tree diagrams are extended in grade 7 where students, in the context of an investigation of traffic patterns, learn to understand probability as a numerical measure calculated from a ratio. In grade 8, probability is formalized, and students are introduced to simple rules that help define ways to think about quantifying the probability of a future event.

Measures of Uncertainty

Students initially consider such questions as, "On which floor is it likely that Frog Newton landed?" (Fig. 2.6). They intuitively know there is a degree of uncertainty involved.

Statistical inference and probability are brought together in grade 8, where students predict which of the given samples come from a particular population. In the investigation, they learn to quantify the uncertainty involved in the sampling process. By confronting issues such as the difference between the local and the national mortality rates on cancer, they learn about the uncertainty inherent in any sample and how that uncertainty can affect decisions.

Simulation is a powerful tool for learning to understand a probabilistic situation, and one that is used in the probability strand. Students use a variety of tools, from number cubes, coins, and random number tables to computer- or calculator-generated random numbers, to simulate different probability situations throughout the probability strand.

In addition, it should be noted that *Mathematics in Context* units have been developed around interesting and relevant contexts. Every unit has connections to content outside the mathematics classroom. In Picturing Numbers (Boswinkel et al., 1997), students analyze the populations of elephants on the African continent. Growing mung beans is a key activity in Insights Into Data (Wijers, de Lange, Shafer, & Burrill, in press). The environment is the focus of the unit Statistics and the Environment (Jonker, Querelle, et al., 1998). Percolation is a context in Great Expectations (Roodhardt, Wijers, Cole, & Burrill, in press). The unit Digging Numbers (de Lange et al., in press) is centered around the theme of archaeology. Students make connections to history when they create number line plots of the dates states entered the Union and analyze graphs showing the ages of Presidents of the United States at their inaugurations.

Also, there is a multitude of connections between the statistics strand and the other strands, particularly with the number and algebra strands. For example, the concept of average is the key context for ratio in the grade 6 number unit Ratios and Rates (Keijzer, Abels, Brinker, Shew, & Cole, 1998).

HILLARY IS WALKING TO THE SCIENCE LAB CARRYING HER PET BULLFROG, NEWTON.

NEWTON, IN FEAR FOR HIS LIFE...

...JUMPS OUT OF HIS AQUARIUM AND HOPS OFF AS FAST AS HIS LITTLE LEGS CAN CARRY HIM.

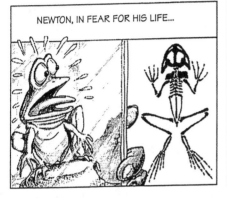

Below you see two tiled floors.
Hillary finally caught Newton. He was sitting on this tile: ➡️

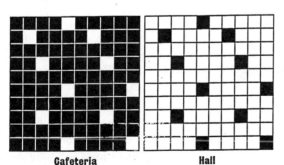

Cafeteria   **Hall**

6. On which of the two tiled floors do you think Newton was hopping? Explain.

7. What if, instead, Newton was sitting on this tile:

Is it likely he was on the same floor?

FIG. 2.6. Sample problem from Take A Chance. Reprinted by permission of the Encyclopaedia Britannica Educational Corporation.

A sample survey is used to generate information that is described in percentages. Mean temperatures are a context used with integer operations in the seventh-grade unit Operations (Abels, Wijers, Burrill, Simon, & Cole, 1998). Graphic displays of data are prominent in the algebra units Tracking Graphs (de Jong, Querelle, Meyer, & Simon, 1998) and Ups and Downs (Abels, de Jong, et al., in press), as well as in Building Formulas (Wijers, Roodhardt, et al., in press). Within the statistics and probability units, percents, fractions, ratios, and proportion also occur in a variety of ways to describe data, particularly when students learn to express probabilities. Measurement ideas are reinforced as students collect and analyze measurement data. In several units, students use fractions as slopes to compare measurements. Estimation is a constant theme. For example, students estimate based on numerical data about the environment, and estimate characteristics of a distribution from a graph. The relation of a linear equation to a set of data is explored in grade 8 as students write equations and analyze the slope in the terms of a data set, and inequalities are investigated in several units.

Summary

By working through these units we expect students to learn important ideas about statistics and probability. Students are given problems situated in real contexts and encouraged to mathematize the problem (going from informal ideas to formal mathematical terms and symbols). In MiC, mathematizing is an iterative process that involves using visualizations, models, and different representations to develop concepts and that includes the interaction and intertwining of ideas from the other strands of mathematics as well as from other disciplines.

SAMPLES OF STUDENT WORK ON MiC UNITS

The following samples of student work have been adapted from the extensive pilot and field testing conducted in schools in several states and in Puerto Rico and illustrate how three of the goals of the strand are realized in the classroom. Data collected during the testing were used to revise the units for commercial publication.

One goal is that students will be able to make conclusions based on data and descriptions of data and create an argument to justify their conclusions.

To make conclusions and create arguments, students need to be able to communicate in a logical and coherent way the conclusions and decisions they make based on data. Initially most students write in general terms, using very few numbers, mathematical facts, or graphic representations in their discussion. In Picturing Numbers (Boswinkel et al., 1997) at grade 5, two students describe what each of two graphs (Fig. 2.7) tell them:

Ricy: The first one shows you use alot less gas in 1950 than in 1989. In the first one it looks like they used wood to make a fire more than they used gas to start one. In 1989 they used a lot more electricity. Both in 1950 and in 1989 fuel/oil was used about the same. Also in 1950 and in 1989 other and none is about the same.

Tierza: The graphs show the home heating fuel in the U.S. In 1950 more people used coal than in 1989. Gas has taken over very much of the products in 1950 from 1989, from less than half to more than half. Electricity became more of usage because electricity costs and today it costs too but now almost a fourth of the graph shows that electricity is used. Fuel oil in 1989 is about 1/6 of the graph, but it used to be almost a quarter.

Ricy does not set the contextual frame for the story—the big picture the graphs are supposed to represent. He uses "alot," "about the same," "a lot more," rather than specific comparisons. Tierza's description demonstrates a different understanding about communicating and building arguments. Although her grammar is not quite what it should be, she explains the problem context and has begun to use mathematics to make comparisons,

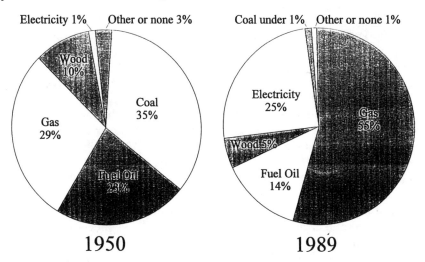

Note: Percentages do not add to 100 due to rounding.

FIG. 2.7. Heating the United States: comparing pie graphs (adapted student work). Source: U.S. Bureau of the Census, Statistical Brief SB/91, August 1991. Reprinted by permission of the Encyclopaedia Britannica Educational Corporation.

blending numbers and words together in her explanation. She also relates the mathematics to the graphs.

Students continue to learn to discuss and write about their conclusions and observations in other statistics and probability units and in units from the other content strands. In grade 6 in Dealing With Data (de Jong, Wijers, Middleton, Simon, & Burill, 1998), they consider the problem of making sense out of large quantities of data by using number summaries and graphs. One activity examines the ages of U.S. Presidents at their inauguration (Fig. 2.8).

Kuong

Demicrats and Republicans

   This report is about demicrats and republicans. The demicrats number is not even on the stem-and-leaf graph. The mean was 52.3 and the median was 54 and the mode was 57. Further down on the page, the box plot has a lower quartile of 49 and an upper quartile of 57. The box plot goes from 40 to 65. The highest it goes is to 65 and the lowest is to 42. They all make 25%. The graph on the next page shows 12 presidents are from 50 to 60. Next I'm going to talk about the Republicans. The republicans are pretty much like the demicrats but not in all ways. So theres not very much to say.

The work done by Kuong illustrates a swing to the other direction: Instead of using general terms, he has done all the calculations and reports them without any contextual relation. Note that Kuong's graphs, although carefully and accurately constructed, are without referent labels: They could represent anything at all about "Demicrats," from number of days in Congress, to number of votes, to their ages. Kuong's summary numbers are correctly calculated, and he displays an understanding of what the numbers mean in a context. Is the mode an accurate measure of a typical value in this case? Are median and mean meaningful in the situation? When questioned, Kuong revealed he was comparing the inaugural ages of Presidents from the two political parties in the United States. He makes no attempt to show the data in a way that could facilitate comparison, such as in a back-to-back stem-and-leaf plot or two box plots on the same scale. His descriptions deal with only one set of data at a time, and although he follows procedures, he does not succeed in using statistics to make sense of the data.

As students develop in their ability to communicate, they are asked to read and analyze statistics and data and how these are used in the media. As an assessment in grade 8 Insights Into Data (Wijers, de Lange, et al., in press), pilot students were asked to find a graph in the newspaper and write a critical analysis of the information it contained (Fig. 2.9). Shannon's report, although not exemplary, demonstrates her understanding of how to organize her writing: problem statement, numerical summaries, usefulness of information,

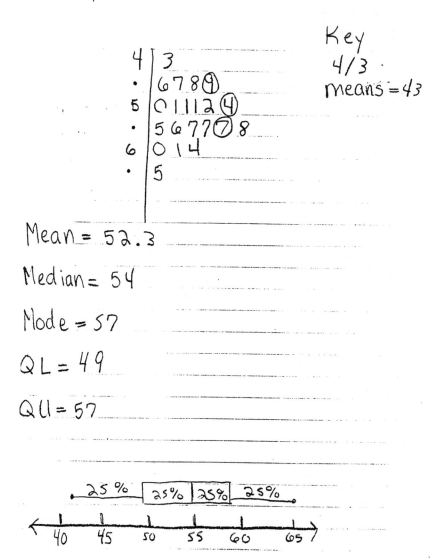

Demicrats

Key
4/3 ·
means = 43

```
4 | 3
  . | 6 7 8 ⑨
5 | 0 1 1 1 2 ④
  . | 5 6 7 7 ⑦ 8
6 | 0 1 4
  . | 5
```

Mean = 52.3

Median = 54

Mode = 57

QL = 49

QU = 57

25% 25% 25% 25%

40   45   50   55   60   65

FIG. 2.8. Inaugural age: using graphs to show mean, mode, and median (adapted student work). Reprinted by permission of the Encyclopaedia Britannica Educational Corporation.

VOX POP

Which would you prefer as a name for your race?

	FEB. 1989	FEB. 1994
AFRICAN AMERICAN	26%	53%
BLACK	61%	36%

From a telephone poll of 503 African Americans taken for TIME/CNN on Feb 16-17 by Yankalovich Partners Inc. Sampling error ± 4.5% "Not sures" omitted

FIG. 2.9. African American or Black: describing a survey (adapted student work). Reprinted by permission of the Encyclopaedia Britannica Educational Corporation.

and comments. The report also displays her ability to think critically about the article and integrate numbers in a meaningful way in her description.

Shannon

"Which would you prefer as a name for your race?"

Time/CNN issued this survey that was taken by African American people on February 16–17, 1994. The people had two choices. The results were that in 1989 26% preferred to be called African American, and 61% preferred to be called Black and in 1994 53% preferred to be called African American and 36% preferred to be called Black. This information helps the people who talk to African Americans, and so that people of other races don't offend them. It also shows that 27% of the people changed their minds about what is the best name to use or else new people moved in or maybe they surveyed a very different sample. There were some biases such as it was a telephone poll, and there was only 503 people surveyed. I don't know if those 503 people were all over the USA and the results of the survey was only in Time magazine.

A second goal is that students will be able to set up a data collection experiment or survey. They will be able to describe data visually and numerically. In the following example from Insights Into Data (Wijers, de Lange, et al., in press), grade 8 students gain experience in designing an experiment and develop their understanding of graphs and how graphs can be useful in conveying information. Students tracked the growth of mung bean sprouts over a 7-day period and produced displays of their results. The beans were grown in four different solutions (water, Coca-Cola, salt, and Sprite), with each group of students assigned to use one of the solutions (Fig. 2.10). The class had learned how to make box plots, stem-and-leaf plots, and number-line plots. They had not learned about plots over time nor how to handle multiple measurements. Their task was to summarize and illustrate the length of sprouts grown in their solutions and to show how the sprouts in a given solution grew over the time period. The groups reported their results in an oral presentation to the class. To compare the solutions, each group made box plots on a number line on the board. They also recorded the final length of each bean on notes, color-coded per solution, and pasted the notes on a number line plot on the board (Fig. 2.11). Students used graphs to summarize the growth length of sprouts and to indicate what different graphs conveyed about the length. Those who used the salt solution struggled more than the others to find a way to describe their growth graphically. It was easy to report orally that their beans didn't sprout, but it was not easy to convey that message using a number plot. One group resorted to making up growth lengths in order to think about how to analyze concrete lengths. They then applied the same techniques with all of their zeros to make their plots.

To address the problem of how the lengths changed over time, students were forced to create new procedures (Fig. 2.12).

Students recorded the length of each sprout each day. From the graph, it isn't possible to see what happened to an individual sprout, but only to the group as a whole. The advantage of this graph is that it reveals the variability in growth as well as the trend. (By day 7, the variability had grown from a range of 8 mm to 59 mm.) Other students summarized each day's growth with a mean or median (day, median growth), which displayed the trend but not the variability. Students learned to understand what is lost in the use of summary statistics.

The graph in Fig. 2.12b shows what happened to each individual sprout over the 7-day period. One sprout grew tremendously over the weekend, which vividly demonstrated for students the danger of extrapolation. What if the experiment had stopped on day 5? Two other sprouts grew less over the entire weekend than on any one of the preceding days. Students had many questions about the differences in growth, and possible causes, exemplifying the critical thinking aspect of the curriculum as well as the notion that statistical studies often raise more questions than they answer.

Note the diversity in the strategies and the rich mathematical understanding that can be observed from the strategies used. These results revealed that, left to themselves with a problem and some basic information, students can find very appropriate and diverse solutions. This illustrates a fundamental pedagogical dilemma: how to provide students with a standard set of tools

Final Length

```
   15    25    35    45    55    65    75
    |-----|-----|-----|-----|-----|-----|-----
   x      xxxx  xxxxx xxxxx  xxx x              x
          xxxx  xxxxx xxxx
          xxxx  xxxx   x
          xxxx  xxxxx x
                 xx
                  x
```

Max = 75 mm

R = 67 mm

Stem and Leaf for Final Length

```
1 | 8
2 | 45555678
3 | 00024455555578
4 | 001245 5577
5 | 0033358
6 | 0
7 | 5
```

DAY 1

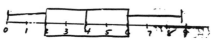

DAY 7

Mung Bean — Grown with tap water — Sprouts

Length (mm)

Leaves

FIG. 2.10. *(Continued)*

50

BEAN SPROUTS

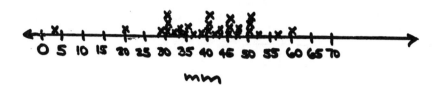

Min = 3
Max = 60
Range = 57

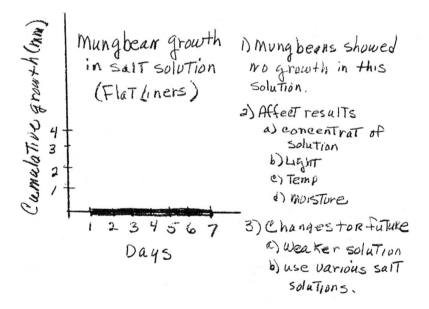

Mungbean growth
in salT soluTion
(FlaT liners)

1) Mung beans showed
no growth in this
soluTion.

2) AffecT resulTs
   a) concenTraT of
      soluTion
   b) LighT
   c) Temp
   d) moisTure

3) Changes for fuTure
   a) weaker soluTion
   b) use various salT
      soluTions.

FIG. 2.10. Mung bean sprouts: using graphs to show comparisons (adapted student work). Reprinted by permission of the Encyclopaedia Britannica Educational Corporation.

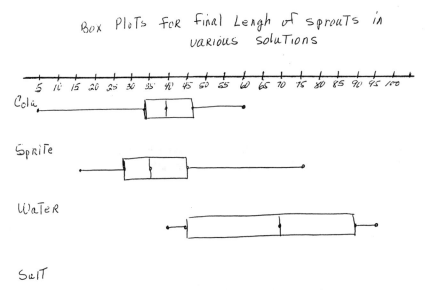

FIG. 2.11. Student presentation of final lengths (adapted student work). Reprinted by permission of the Encyclopaedia Britannica Educational Corporation.

without stifling their creativity and their own instincts about solution strategies. A third goal is that students will be able to use their knowledge of probability to make predictions about unknown events.

Making decisions based on surveys and samples is a recurring theme in the unit. Students are informally introduced to the sampling process in the early probability and statistics units when they are asked to read and evaluate surveys from the media, as well as to construct their own surveys. Tim, grade 7, created the table shown in Fig. 2.13 and described his results:

Question: Which brand name of cola do you prefer to drink, Pepsi or Coke?

I asked people, family, friends, and students, of every age and of both sexes male and female which cola they prefer to drink. When I finished my survey I decided the best way to show the results was a bar graph. I interviewed a total of 71 people, 51 kids and 20 adults. The results were, a total of 47 people preferred Pepsi (33 kids and 14 adults) and 24 people preferred Coke (18 kids and 6 adults). It doesn't look like there is any connection between being a kid or an adult and liking one or the other. Most people liked Pepsi. This survey was good for choosing what kind of pop sells. There were many different reasons why people made their choices.

Pepsi: Like the slogan, it just tastes better, all my friends drink it, the can looks better.

Coke: It tastes better, I think old people only drink Pepsi, it's refreshing, it quenches my thirst after playing sports.

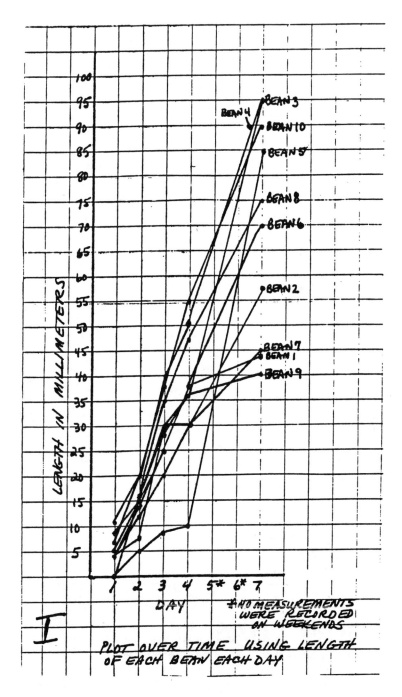

FIG. 2.12. *(Continued)*

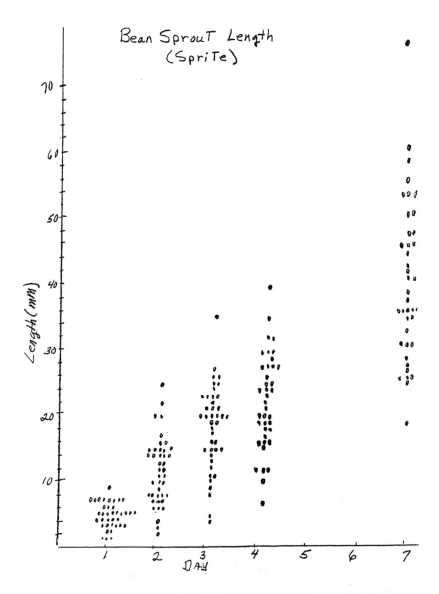

FIG. 2.12. Growth over time: using graphs to show growth (adapted student work). Reprinted by permission of the Encyclopaedia Britannica Educational Corporation.

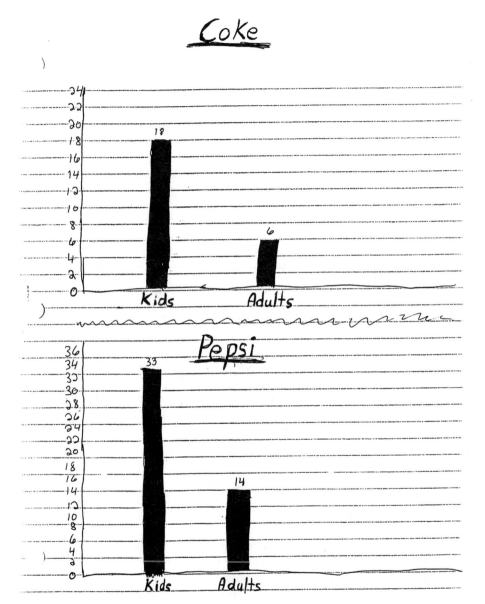

FIG. 2.13. *(Continued)*

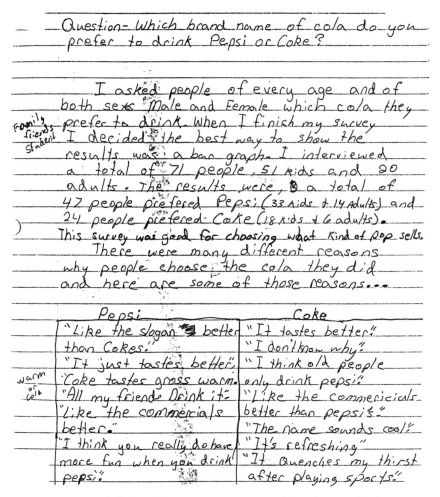

Question— Which brand name of cola do you prefer to drink Pepsi or Coke?

I asked people of every age and of both sexs Male and Female which cola they prefer to drink. When I finish my survey I decided the best way to show the results was a bar graph. I interviewed a total of 71 people, 51 kids and 20 adults. The results were, a total of 47 people prefered Pepsi (33 kids + 14 Adults) and 24 people prefered Coke (18 kids + 6 adults). This survey was good for choosing what kind of pop sells. There were many different reasons why people choose the cola they did and here are some of those reasons...

Pepsi	Coke
"Like the slogan better than Cokes."	"It tastes better."
	"I don't know why."
"It just tastes better."	"I think old people only drink pepsi."
"Coke tastes gross warm."	
"All my friends Drink it."	"Like the commericials better than pepsi's."
"Like the commercials better."	"The name sounds cool."
"I think you really do have more fun when you drink pepsi."	"It's refreshing."
	"It Quenches my thirst after playing sports."

FIG. 2.13. Soft drink preferences: using a two-way table to show results of a survey (adapted student work). Reprinted by permission of the Encyclopaedia Britannica Educational Corporation.

In Great Expectations (Roodhardt et al., in press), students extend their understanding of the "connection" between two results as they investigate independent and dependent events. Using a two-way table, they formalize a procedure to deal with unequal bases and dependent events as they quantify the information (Table 2.2).

If there is no connection between the events "adult/child" and "soft drink preference," that is, if the events are independent, then the ratios for each column in the table will be almost the same. If there is a connection, the ratios will be different. In this case (14/47 = 30% and 6/24 = 25%), the ratios are relatively close, and there seems to be no connection between the events.

TABLE 2.2
Survey of Cola Preference Results (Adapted Student Work)

	Prefer Pepsi	Prefer Coke	Total
Adults	14	6	20
Kids	33	18	51

Preferring Pepsi does not seem to be a function of being either an adult or a child. The question of whether a 5% difference is large enough to be significant and how this difference can vary depending on sampling is addressed in the unit at a very informal level, but deeply enough to raise issues and begin laying the foundation for more formal inferential techniques.

Summary

In today's information society, students must learn to deal with data and learn to use data to solve problems. *Mathematics in Context* is one example of an instructional program designed to encourage students to acquire a critical attitude about the role of statistics and probability in our society. Through a carefully designed set of activities, students have the opportunity to reinvent important statistical and probabilistic concepts and procedures as they shift from informal to formal conceptualizations. The examples of student work from some of the units show that students can use statistics to deal with data. They can collect information, judge the appropriateness of a sample of data, picture the properties of batches of data, reason from that information, and draw conclusions based on that data. In fact, students find such activities engaging and often demonstrate very creative ways of dealing with information.

The activities in the statistics and probability strand are diverse and provide students with an introduction to major concepts. Pilot- and field-test evidence indicates that students seem to be finding the materials interesting and useful. The final test of how effective the material will be, however, is whether students will recognize opportunities to use statistical procedures in problems other than those in the statistics/probability units.

Making such activities an integral part of a middle-school curriculum creates pedagogical problems. Too many teachers, for example, have little or no background in statistics and probability. The process of gathering, summarizing, and interpreting data is time-consuming and involves quite different teacher–student interactions than have been common in most classrooms. Interesting, good, well-designed activities, such as those in *Mathematics in Context,* alone are not sufficient. For statistics and probability to be taught so that all students understand and can use the concepts and procedures to solve problems, the units need to be integrated into the total curriculum and be considered as important to teach as are other topics in mathematics.

REFERENCES

Abels, M., de Jong, J., Meyer, M., Shew, J., Burrill, G., & Simon, A. (in press). Ups and downs. In National Center for Research in Mathematical Sciences Education and Freudenthal Institute (Eds.), *Mathematics in context: A connected curriculum for grades 5–8*. Chicago: Encyclopaedia Britannica Educational Corporation.

Abels, M., Wijers, M., Burrill, G., Simon, A., & Cole, B. (1998). Operations. In National Center for Research in Mathematical Sciences Education and Freudenthal Institute (Eds.), *Mathematics in context: A connected curriculum for grades 5–8*. Chicago: Encyclopaedia Britannica Educational Corporation.

Boswinkel, N., Niehaus, J., Gravemeijer, K., Middleton, J., Spence, M., Burrill, G., & Milinkovic, J. (1997). Picturing numbers. In National Center for Research in Mathematical Sciences Education and Freudenthal Institute (Eds.), *Mathematics in context: A connected curriculum for grades 5–8*. Chicago: Encyclopaedia Britannica Educational Corporation.

de Jong, J., Querelle, N., Meyer, M., & Simon, A. (1998). Tracking graphs. In National Center for Research in Mathematical Sciences Education and Freudenthal Institute (Eds.), *Mathematics in context: A connected curriculum for grades 5–8*. Chicago: Encyclopaedia Britannica Educational Corporation.

de Jong, J., Wijers, M., Middleton, J., Simon, A., & Burrill, G. (1998). Dealing with data. In National Center for Research in Mathematical Sciences Education and Freudenthal Institute (Eds.), *Mathematics in context: A connected curriculum for grades 5–8*. Chicago: Encyclopaedia Britannica Educational Corporation.

de Lange, J., Roodhardt, A., Pligge, M., Simon, A., Middleton, J., & Cole, B. (in press). Digging numbers. In National Center for Research in Mathematical Sciences Education and Freudenthal Institute (Eds.), *Mathematics in context: A connected curriculum for grades 5–8*. Chicago: Encyclopaedia Britannica Educational Corporation.

Gravemeijer, K. (1990). Context problems and realistic mathematics instruction. In K. Gravemeijer, M. Van den Heuvel, & L. Streefland (Eds.), *Contexts, free productions, tests, and geometry in realistic mathematics education* (pp. 10–32). Utrecht, The Netherlands: OW & OC Research Group.

Jonker, V., Querelle, N., Wijers, M., de Wild, A., Spence, M., Fix, M., Shafer, M., & Browne, J. (1998). Statistics and the environment. In National Center for Research in Mathematical Sciences Education and Freudenthal Institute (Eds.), *Mathematics in context: A connected curriculum for grades 5–8*. Chicago: Encyclopaedia Britannica Educational Corporation.

Jonker, V., van Galen, F., Boswinkel, N., Wijers, M., Simon, A., Burrill, G., & Middleton, J. (1997). Take a chance. In National Center for Research in Mathematical Sciences Education and Freudenthal Institute (Eds.), *Mathematics in context: A connected curriculum for grades 5–8*. Chicago: Encyclopaedia Britannica Educational Corporation.

Keijzer, R., Abels, M., Brinker, L., Shew, J., & Cole, B. (1998). Ratios and rates. In National Center for Research in Mathematical Sciences Education and Freudenthal Institute (Eds.), *Mathematics in context: A connected curriculum for grades 5–8*. Chicago: Encyclopaedia Britannica Educational Corporation.

National Center for Research in Mathematical Sciences Education and Freudenthal Institute. (Eds.). (in press). *Mathematics in context: A connected curriculum for grades 5–8*. Chicago: Encyclopaedia Britannica Educational Corporation.

National Council of Teachers of Mathematics. (1989). *Curriculum and evaluation standards for school mathematics.* Reston, VA: Author.

Roodhardt, A., Wijers, M., Cole, B., & Burrill, G. (in press). Great expectations. In National Center for Research in Mathematical Sciences Education and Freudenthal Institute (Eds.), *Mathematics in context: A connected curriculum for grades 5–8.* Chicago: Encyclopaedia Britannica Educational Corporation.

Wijers, M., de Lange, J., Shafer, M., & Burrill, G. (in press). Insights into data. In National Center for Research in Mathematical Sciences Education and Freudenthal Institute (Eds.), *Mathematics in context: A connected curriculum for grades 5–8.* Chicago: Encyclopaedia Britannica Educational Corporation.

Wijers, M., Roodhardt, A., van Reeuwijk, M., Burrill, G., Cole, B., & Pligge, M. (in press). Building formulas. In National Center for Research in Mathematical Sciences Education and Freudenthal Institute (Eds.), *Mathematics in context: A connected curriculum for grades 5–8.* Chicago: Encyclopaedia Britannica Educational Corporation.

PART TWO

TEACHING STATISTICS

CHAPTER THREE

Graphical Representations: Helping Students Interpret Data

George W. Bright
University of North Carolina at Greensboro

Susan N. Friel
University of North Carolina at Chapel Hill

Over the past decade, documents on mathematics education reform (e.g., National Council of Teachers of Mathematics [NCTM], 1989) have highlighted the importance of including statistics and data analysis throughout the school mathematics curriculum. For many mathematics teachers, however, statistics continues to be a content area with which they have little experience. This may be due in part to the fact that mathematics is often taught as a discipline focused on procedures, yet only small parts of current conceptualizations of statistics fit within a procedural paradigm. The "reform view" of statistics demands that instruction be organized around critical inquiry rather than mastery of specified procedures. Statistical thinking provides a process for understanding the world. Those who engage in statistical thinking are engaged in making sense of their world, an activity that has considerable payoff not only in school mathematics but also in the world outside of school.

Statistics as mathematics content involves developing data sense as well as familiarity with a body of tools for making sense of data. Data sense includes being comfortable with formulating questions as well as gathering and interpreting data in ways that respond to those questions. However, data sense extends well beyond this to include comfort and competence (e.g., aspects of numeracy) in reading and/or listening to reports based in statistics found in newspapers, magazines, television, and other forms of popular press (Joram, Resnick, & Gabriele, 1995). Reading and listening using the lens of statistics is more than understanding the graphs and statistics that are presented. It involves being able to make translations among various representations of data and requires evaluation of the statistical investigation process used.

What make a problem a statistical investigation are the way the question is posed, the nature of the data and the ways the data are collected, how the data are examined, and the types of interpretations made from such examination. These four components have been outlined by Graham (1987). Kader and Perry (1994) suggested a fifth phase—namely, communicating the results of the interpretation (Fig. 3.1). The main use that we have made of the model is to give structure and direction to our understanding of the types of reasoning used in statistical problem solving. We had originally viewed communication as mainly part of the interpretation phase, but now it seems useful to separate out communication as an overriding element, because by doing so, we emphasize the importance of utility in statistical activity. That is, statistical understanding is not useful unless that understanding can be communicated to others.

This "process notion" of statistical investigations is consistent with the NCTM discussion of statistics as a standard of the curriculum across grades K–12 (NCTM, 1989). The intent of the diagram in Fig. 3.1, however, is to portray the elements of statistical understanding as a dynamic process, with the knowledge that the cycle might need to be repeated several times to truly understand a real-world situation or that the cycle itself may need to be interrupted so that an earlier phase can be clarified (e.g., to clarify the question). The remainder of this chapter deals with the analysis phase of the process, during which data are organized into graphs. We ask the reader, however, to always keep in mind that the use of graphs is never an end in itself. Rather, graphs are only one way to organize information meaningfully as part of the process of statistical investigation.

BAR GRAPHS FOR DATA IN RAW AND REDUCED FORMS

When students engage in a statistical investigation, they collect data about the individual elements in a sample. An initial step in the analysis phase often involves making one or more representations of these data. This step may involve a decision about whether the data will be left as "ungrouped" data (raw data) representing individual elements of a sample or will be grouped in some way (reduced data). Data reduction can occur either by grouping the data values themselves (e.g., number of times each data value occurs) or by grouping the data according to attributes of the elements in the sample (e.g., gender). There are a variety of ways to represent data, and students may experiment with several in the course of examining a data set. Most choices require either implicit or explicit attention to data reduction: Do I use a line plot? a bar graph? a stem plot? a histogram? The way data are reduced is intimately related to the types of representations used.

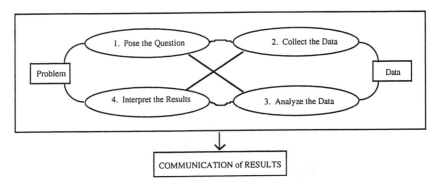

FIG. 3.1. The process of a statistical investigation.

The decision about grouping has implications for the way that a representation of data is made and interpreted. When data are ungrouped, each data value has its own plot element (e.g., bar). When the data come from students in the class, each student can identify the bar that belongs to herself or himself. The unit of measure for "sizing" each bar is related to the unit of measure for the values themselves. For example, a value of 10 is represented with a bar that is twice as tall as a bar showing a value of 5.

Once data are grouped, however, the individual data values are "buried" within plot elements. For example, in a bar graph, all of the values of 10 may be represented collectively in a single bar. The size of the bar is related to the number of 10s that are in the data set. If the number of 5s and the number of 10s are the same, the bars for showing the frequencies of these data values are the same height, even though the data values themselves are not the same. Once data reduction is used, each data value is associated with a common-sized unit (e.g., an X on a line plot or a bar showing frequency on a bar graph). The unit of measure for sizing the plot elements is unrelated to the unit of measure for the data values.

The role of the axes in bar graphs depends on whether the data are ungrouped or grouped. For raw data, the x-axis is a vehicle for labeling the data values that name the bars. The y-axis provides the actual data values when matched with the heights of the bars. When data are reduced with simple tallying, the x-axis is used to provide the actual data values. Each of these values may have a bar associated with it. The y-axis provides the frequency count of the occurrence of each of these values when matched with the heights of each bar (Fig. 3.2). It seems critical to observe that the roles of the y-axis for raw data and the x-axis for reduced data serve the same purpose: that is, providing information about the actual data values. If learners are to relate representations showing raw and reduced data, they need to be aware of the change of perspective required as these representations change.

UNGROUPED DATA

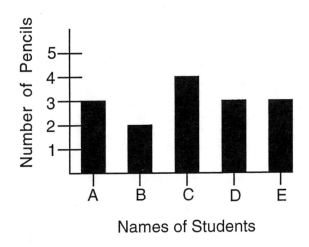

GROUPED DATA

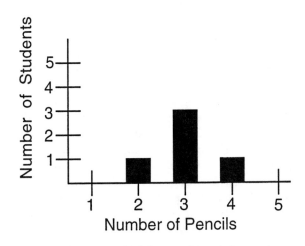

FIG. 3.2. Distinguishing the use of axes in bar graphs.

Tables seem frequently to be used somewhat haphazardly in instructional programs that involve working with data. However, with the potential for confusion that exists in representing data, tables may play an important role as an intervening representation that can smooth the transition between representing raw and reduced data. That is, tallying may help make the distinction between ungrouped and grouped data clearer, because creating tallies of data values seems to require learners to focus on the relationships between data values and their frequencies. Tallying may also be useful in developing a sense of reversibility in data representations. That is, although tallies are usually made from raw data, students may also be asked whether they can "reconstruct" the original data set from graphical display(s). With line plots and bar graphs, for example, this is possible, whereas with histograms and box plots it is not possible. This may help students develop an awareness and understanding of the subtleties that serve to distinguish one representation from another. As noted by Hancock, Kaput, and Goldsmith (1992), technology may be a particularly effective instructional vehicle for helping students relate representations, but more study of this issue is required.

In the next few sections of this chapter these ideas are illustrated. In each section, the representations to be compared are discussed, interview questions that were asked of students are presented, and sample responses from student interviews are quoted. These responses illustrate the ways in which students think about the pairs of representations and begin to clarify what students understand and what they misunderstand.

The pairs of representations that are discussed were selected for a number of reasons. First, the representations individually are common in the school mathematics curriculum. Yet it seems clear to us that teaching them in isolation from each other does not help students develop a deep understanding of the ways they may communicate about data. Second, the transition from ungrouped or raw data displayed on bar graphs to grouped data displayed on line plots, bar graphs, stem-and-leaf plots, and histograms seems to be an appropriate sequence for helping students understand strategies for reducing increasingly messy and complex data. For example, data about family size (with a usual range of about 10 people) are efficiently represented in a standard line plot or bar graph, data about breath-holding (with a usual range of about 100 seconds) are better represented in a stem-and-leaf plot than in a bar graph, and data about number of minutes students spend watching television in an evening (with a usual range of several hundred minutes) are better represented using a histogram than a stem plot. Each of these representations responds to the need to identify patterns as the data exhibit greater spread and/or variability. Third, there are mathematical and conceptual connections between the pairs of representations discussed. Helping students understand these connections seems to be one way to help deepen their understanding of statistics. The choice of a representation that "best" communicates the characteristics of a particular data set, however, must respond to the nature of that data set. Instruction should help students understand the types of data for which particular representations are most applicable.

OVERVIEW OF THE STUDY

During the fall of 1994, we conducted a study of the ways that students in grades 6, 7, and 8 make sense of information in graphical representations and make connections between pairs of graphs. Students were tested both before and after an instructional unit developed specifically to highlight connections between pairs of graphs. Four students from grades 6 and 8 were also individually interviewed before and after the unit. Interviews were tape-recorded and transcribed. (See Friel & Bright, 1995b, for the tests, interview protocols, and instructional unit used.) The contexts used in the interview problems were intended to be easily understood. Further, they matched the contexts used in instruction, although the specific questions asked in the interviews were not included in the instructional unit. We hoped in this way to prevent students from being unable to demonstrate their understanding simply because they did not understand the contexts of the problems.

Data from the interviews of eighth-grade students are used here to illustrate the difficulties and successes that students experienced in attempting to understand the material from this unit. These responses seem to cover the range of responses of all students who were interviewed, but the eighth-grade students were better able than the sixth-grade students to explain in words what they were doing. The interview excerpts hopefully illustrate students' understanding about graphical representations. The data from the written tests also seemed generally to support the observations made about the interview responses (Friel & Bright, 1995a), but lack of space here prevents discussion of specific written responses.

BAR GRAPHS FOR UNGROUPED AND GROUPED DATA

It is not uncommon to find activities in elementary-school mathematics textbooks in which students gather data and represent their personal information either with towers of interlocking cubes or as individual bars on a graph. Putting these towers or bars together creates an "ungrouped data" representation. Each student can identify her or his own data value by pointing to a particular bar. The process of translating this ungrouped bar graph to a "standard" bar graph, however, is not necessarily easy. Students have to conceptualize the tallying of values and shift their view of the appropriate unit for "sizing" the plot elements. The first and third graphs in Fig. 3.3 are examples of these two kinds of bar graphs showing numbers of children in families.

Interview Responses About Bar Graphs

The interview questions are given in Fig. 3.3. Excerpts from responses to the interview questions are presented in Figs. 3.4 and 3.5. Questions call explicit attention to the differences in the ways that data are presented in these representations. By being asked the same question across representations (e.g., Can you use the graph to find out how many children are in all the families in the class all together?), students must decide not only what information is presented in each representation but also whether identical information can be extracted from the different representations.

One confusion about the role of labels is revealed in the responses of students during the preinstruction interviews. In response to the questions about the meaning of the bars, student T seemed to answer correctly for the first bar but not for the second bar, students A and G seemed to reverse the roles of the axes for both bars, and student J seemed to interpret the axes correctly. After instruction, only student A continued to reverse the meaning of the axes. This suggests that students do improve their skill at reading bar graphs, although the confusion of individual students may be difficult to overcome.

A second confusion about how to process information is revealed in the responses to questions about the number of students in the class and about the total number of children in all the families of children in the class. When asked how many students are in the class, prior to instruction student J seemed confused about the context of the question, but after instruction all students seemed able to understand the question, although student A identified the wrong information from the graph to use in answering the question. Students seemed to approach the task slightly more directly after instruction than before instruction.

Both before and after the instructional unit, two students (students G and J before instruction and students T and J after instruction) were able to determine how many children are in all the families of all the students in the class, or to specify a correct process for finding this number. However, after instruction the information that was identified by even the incorrectly responding students as important for answering this question was more relevant to the question. Even after instruction, students A and G seemed confused about whether to find the total number of students in the class or the total number of children in all the families, even though student G had correctly found the number of students in the class. It is not clear why students' processing of information goes astray, even when the context is understood and relevant information can be read off the graph.

Read the context of the problem situation to the subject.

A class of students has been collecting information about themselves. One question that they wanted to find out was how many children each person in class had in his/her family.

1. The students made a graph to show the information they collected. This is their graph (show graph of unordered data). What does that graph show?

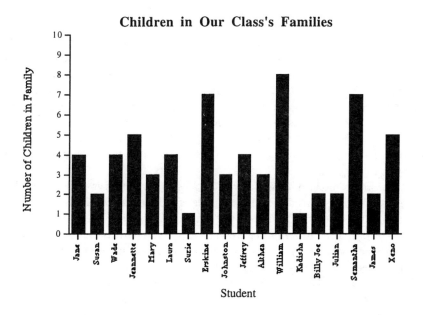

2. The students in the class decided that they would redo the graph and make it a little different, so they made this graph. (Show graph of ordered data.) How are these two graphs alike and different?
3. How many students are in the class? How did you find your answer?
4. Can you use the graphs to find out how many children are in all the families in the class all together? Have subject provide explanation.

FIG. 3.3. *(Continued)*

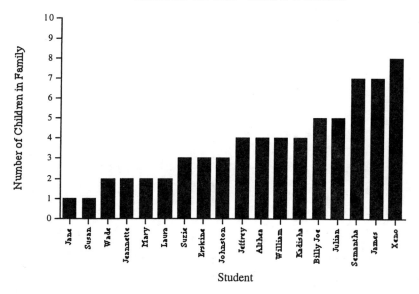

Children in Our Class's Families

(Put first two graphs aside but you may leave them out where they can be seen by subject). Next, their teacher gave them a new graph. (Show bar graph.) She said, "I used the information to make a different kind of graph."

5. Study this new graph for a few minutes.
    a. What does this graph show about the number of children in families? How might the teacher have used the information on the second graph we looked at to help make this graph?
    b. From the information on this graph, describe how you find out how many students are in the class. (Have the subject do the calculations.)
    c. From the information on this graph, describe how you would find out the total number of children that are in *all* the families in the class. (Have the subject do the calculations.)

Children in Our Class's Families

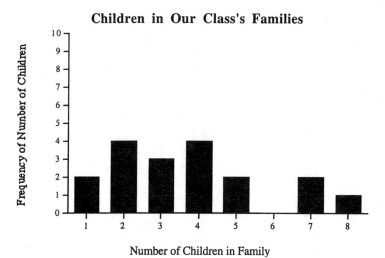

FIG. 3.3. Interview questions to probe ungrouped versus standard bar graph.

Question	Student T	Student A	Student G	Student J
What does the first bar tell you? What does the second bar tell you?	That it's got children in the family. One child in the family—that'd be two families have that. [The second bar tells you] there are four in the family and four children have that kind of family.	That there is one family that has two children. . . . [The second bar shows] two families that have four children in all.	It shows you that someone—two of the—I mean—one of the children has two people in the family. . . . [The second bar] tells you—children have four people in their family.	[The second bar tells you] that two children showed up four times.
From the information on this graph, how can you find out how many students are in the class?	Take the number over here, like 2, and there's 4 and 3 and 4 and 2 and 2 again and 1 and add them all up.	You would say like this one [first bar] is barred as two and this one [second bar] is barred as four and this one [third bar] is barred as three and so on with the rest of them. And you add them, like this one would be two and that one would be four and you'd add them up for the rest of them.	18. We find out the students for every graph [i.e., bar] at the end. You look over and see the number of students that are in there [i.e., each bar] and I looked all over and I found 18. . . . [I added] 2, 4, 3, 4, 2, 2, 1.	I can't tell you how many students there are in the class] cause its got the people in the class and the amount of children in their family. . . . Because that shows how many times there was one child, and the children have brothers and sisters so you can't tell.
From the information on this graph, how can you find out how many children there are in all the families of all the students in the class?	Add up all the bottom numbers. . . . [The sum is] 36. . . . [The answers are not the same because] I might have done that [the answer to question 4] differently than what I did that [the answer to question 5c].	You would add them numbers [the heights of the bars] up and then take the bottom numbers right here [the numbers labeling the x-axis] and you would add them with—you would add them numbers, in other words 18, to the numbers right there [the x-axis labels].	I got 51. . . . I knew there were 18 students so they might have counted themself in their family. And down here [x-axis] it shows me how many people they have in their family. Wait a minute. . . . I don't know if this is right. I got 67. . . . One person in their family, right here, number one. It shows how many students down here [x-axis]. I just times these—2 times 1 and I got 2, so I did the same thing with number two—2 times 4 equals 8. And I did the same thing—times them. Then after I timesed them, I added them up.	You can add up the . . . Wait a second. . . . You add, like, there'd be two right here because there's—two times there's one child and there'd be 8 right there because 4 times you have 2 children. And then—so there'd be . . . 67 because I multiplied and then I added all these numbers up.

FIG. 3.4. Preinstruction interview excerpts: bar graphs for ungrouped and grouped data.

Question	Student T	Student A	Student G	Student J
What does the first bar tell you? What does the second bar tell you?	[The first bar shows] there's two families that have one children—child.	[The second bar means] that there is two people in her class that have four people in her family—their family.	Two people had one person in their family, four people had two people in their family.	[The second bar] says that there are four families with only two children in the family.
From the information on this graph, how can you find out how many students are in the class?	18. . . . I took—the first bar has two, so two, the second bar had four, 2 plus 4 which is 6, then I took the next bar which had three and 6 plus 3 would be 9, . . . I kept adding however many there was.	36. I counted all the bottom numbers of the graph where it says number of children in family.	18. . . . Everywhere the bar graph ends, it's like one is 2. I take 2 plus 4 right here plus 3 plus 4 again plus 2 plus 2 then plus 1.	You know that you'd add up . . . for each family there's at least one student, so you add up all of the—there's 2 children here . . . and there's 4 students here . . . so it's like 6, 7, 8, 9, . . . 18.
From the information on this graph, how can you find out how many children there are in all the families of all the students in the class?	67. . . . In the first one there's two families that's got one so I went 1 times 2 and that's 2, and the next one has four families of two and 2 times 4 which was 8, then there was three families that had three and I did 3 times 3 and that was 9, . . . and added all that total up. [The 67 means] number of children in the family.	You would take, like, where on the bottom graph it's got like 1, then it goes up to 2, and it's filled in to 2, and you just count, like, that one has 2, and that one has 4, that'd be 6, and that one only has 3 so that'd be 9, and that one has 4, that'd be 13, and that one has 2, that'd be 16, that one has 2 so that'd be 18, and that one only has 1 so that'd be 19. [So all of the children in all of the families would be] 19.	You go 18 students—two people—add these up right here—two—I don't know how you could do that. I don't think you could get a total, cause I don't know how you add this thing up. But it would be more easier to add up on the [ordered bar] graph.	I would take this number times this number . . . 2 two times you have 1 child and . . . 4 times you have 2 children and then I would get 2 here and I'd get 8 here and then I would add these two together . . . 3 times 3 is 9 and then I would add it to these two bars [and so on].

FIG. 3.5. Postinstruction interview excerpts: bar graphs for ungrouped and grouped data.

It seems reasonable to assume that all of these students had been exposed to many different bar graphs during both in-school and out-of-school experiences. In spite of this, however, they were not highly successful at answering questions that required deep understanding of the information represented by the graph. Student A's persistence in focusing on "the bottom numbers" both before and after the instructional unit suggests that some misconceptions may be quite resistant to change. Also, the tendency of the students to want to move quickly to manipulation of information (e.g., when student G said, "I don't know how you add this thing up") suggests that more attention may need to be given to helping students talk about information in a graphical display before students are asked to read between or beyond the data (Curcio, 1989, pp. 5–6).

Students T and J showed the most improvement in understanding from preinstruction to postinstruction. Student G seemed somewhat inconsistent in both interviews, and student A seemed confused in both interviews. Perhaps by focusing instruction more explicitly on understanding the connections between bar graphs that display ungrouped and grouped data, students would be better able to answer questions of reading between and beyond the data.

It should be noted that in the third bar graph of Fig. 3.3 (that is, the "standard" bar graph), the numbers used as labels on the x-axis and y-axis both begin implicitly with 0 and explicitly with 1. There is a possibility that confusion may arise simply because the numbers on those two axes are close in magnitude. If a context had been chosen where the numbers on the x-axis had been considerably larger than the numbers on the y-axis (e.g., counting raisins in snack-sized boxes), there might be less confusion caused by the magnitudes of the numbers.

LINE PLOT TO BAR GRAPH

One of the simplest ways to "sketch out" a data set is to create a line plot: that is, recording an X for each data value above the value which is shown on a number line. (Figure 3.6 contains the interview questions for line plot and shows an example of a line plot.) Only one axis is actually shown in a line plot, even though a y-axis representing a frequency count is implicit in the representation. One of the difficulties that students sometimes encounter in making a line plot is that they do not make the Xs the same size, so it may be difficult visually to decide which of two values is most common in the data set. However, in a line plot, each data value is clearly distinguishable, so as with a bar graph of ungrouped data, a student can point to the X that "belongs to her/him." Students often remember exactly where their values are shown in a line plot; for example, "My value is the third X from the bottom above the 7."

A standard bar graph looks much like a line plot, but there are important differences. First, the frequencies of individual data values are no longer visually separated in a bar graph. Each bar represents a group of identical data values, so there may be some confusion about how any particular value is represented or about how many data values there are. The Xs in a line plot can be counted to determine the number of data values, but in a bar graph it is not as easy to determine the size of the data set. Indeed, if the y-axis is not explicitly provided, the original data values cannot be reconstructed at all. When a y-axis is provided, however, it is possible to recreate a list of data values, though it seems easier to make errors in determining frequencies of values in a bar graph than in a line plot. Second, a bar graph has two axes instead of one. Reading the height of a bar requires that the bar be compared with the y-axis, so if students are not comfortable relating information from two variables simultaneously, they are likely to have some difficulty making sense of the bar graph.

In the instructional unit used in our study, students were asked to interpret line plots and bar graphs individually, but they were also asked to decide how these representations were alike and different. Teachers also asked students to find the mode and range of data presented in each way.

Interview Responses About Line Plots and Bar Graphs

The interview questions that we used are shown in Fig. 3.6. Again, asking the same questions about both representations requires students to compare the ways that identical information can be extracted from each representation. Excerpts from the interviews are presented in Figs. 3.7 and 3.8.

Both prior to and after instruction, students were generally able to read the information in the line plot, although there seemed to be more confusion after

Read the context of the problem situation to the subject.

A class of students was getting ready to take an interest survey. They had to fill in their first and last names on an answer sheet using individual boxes for each of the letters in their names. There were 12 spaces available for students to mark their first and last names (not counting the space between the first and last names). The students decided to collect information on how long each person's name was.

1. Some of the students decided to make a line plot showing their information on lengths of names. (Show line plot.)
   a. Describe what you know about the lengths of their names now?
   b. How many of the students will be able to fit their full first and last names in the boxes provided? How can you tell?

FIG. 3.6. *(Continued)*

Total Letters in First and Last Names of Class

```
                                    X
                            X       X
                            X       X
                    X       X  X  X  X
                    X       X  X  X  X  X
                    X       X  X  X  X  X  X  X
  _____
   1  2  3  4  5  6  7  8  9  10 11 12 13 14 15 16
```

Number of Letters in Name

2. Some of the students made a different kind of graph. (Show bar graph.)
   a. The class looked at both graphs. They wondered if the graphs were really all that different from each other. What do you think?
   b. How many students are there in the class? Describe how can you figure out this information from each of the graphs. (Have students do calculations from each of the graphs.)
3. How many total letters are in all the students' names? Describe how you can figure this out from the information on each of the graphs.

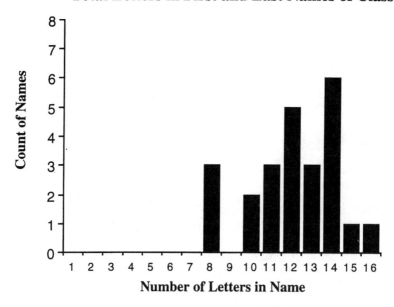

Total Letters in First and Last Names of Class

Number of Letters in Name

FIG. 3.6. Interview questions to probe line plot to bar graph.

Question	Student T	Student A	Student G	Student J
How many students could write their names in a set of twelve boxes?	There's three that's got eight in their names, and 10 have two in—two of them have ten in theirs, and then eleven have three in theirs and twelve have 5 and I add them up, cause 13 or more couldn't fit in them boxes.	I added the numbers that you have—are on the bottom row. You have up to twelve—I mean up to sixteen, but there's not sixteen boxes. There are only twelve so I left 13, 14, 15, and 16, and just added them X's from twelve down.	It's twelve boxes . . . so I looked up the number in the last names. I looked under twelve—I looked under it—five people. And then I looked below [twelve] going down and there were also people that had that too.	Thirteen. . . . There were twelve boxes, so I went ahead and counted these names [the five X's above the 12], and I counted all of them up from eight to twelve.
How many letters are in all of the names for the students in the class?	I'd have to multiply. . . . There's 3 and 8, so 8 times 3 which would equal 24, and there's 2 and 10, so that'd be 10 times 2 which is 20, and then in 11 there's 3, which would be 11 times 3 again which equals 33 [and so on]. You'd add 'em all up together.	You would say, like okay, 8 would represent 3 of the students have 8—8 letters in their name and so that would represent 3 of the students have 8 letters in their name. And 2 of the students have 10, and 3 of the students have 11, and so on. . . . [But] I don't know. I don't understand the question. [Question was repeated.] Probably not. I don't think so.	((The process used was to write down the numbers which were labels on the x-axis and add them up. This student said nothing while carrying out this procedure.))	There's 3 people with 8 so I'd say 8 times 3 is 24, and 10 times 2 is 20, and 11 times 3, and then I would add them all up.

FIG. 3.7. Preinstruction interview excerpts: line plots.

77

Question	Student T	Student A	Student G	Student J
How many letters are in all of the names for the students in the class?	You could add up all the ones that's got numbers on it—letters on it—and like on 8 there was 3 so it'd be 8 three times and that's be like 24. Then 10 had 2 and that was 20, so 24 and 20 would be 44, and you just keep going up and seeing however many there was and multiplying, like the 11 times 3 and that'd be 33, and whatever you got from the last one you'd add 'em all up until you got to the end.	You would take, like, like, like the 8 it shows 3—3 students have 8, so you would say, like, 8 times 3 and then you would get 24. And then, and then like 10 times 2 which would be 20, and then after you got done with that you would add 'em all up, add the numbers all up and you would come out with the total letters.	Add all these things up just times, like 8 times 3 [and so on].	I would get the product, let's say, 8 times 3 is 24, and then here I'd say 10 times 2 is 20, and then after multiplying each bar and getting the product I'd add them all up.

FIG. 3.8. Postinstruction interview excerpts: line plots.

instruction about whether the notion of "writing names in 12 boxes" meant filling all 12 boxes. This apparent confusion may represent an increased sensitivity on the part of students to understanding precisely what a question is asking. Determining how many letters were in all the names for students in the class, however, was a question that was answered much more successfully after instruction than before. This question was designed to be parallel to computing the total number of children from the bar graph (Fig. 3.3), and the universal success with this question in the context of the line plot suggests that the line plot as a representation is simpler than the bar graph. In the line plot each data element is shown (as an X); further there is no y-axis to interpret. Having fewer plot elements to interpret may be one of the reasons for the greater ease in interpreting this information. It also seems important that students could relatively clearly explain their procedures. Multiplication was used frequently in this context, whereas it was used infrequently computing the total number of children. Perhaps the nature of repeating Xs in this representation more clearly suggests multiplication than the area model implicit in each bar in the bar graph. This suggests that students may need to become better able to recognize how mathematical concepts are connected to graphical representations. It may also be important in instruction to be sure that students recognize explicitly that they can identify individual data values quickly in the line plot, but they have to make inferences about individual data values when they are interpreting a bar graph. That is, in a bar graph, individual data values are "buried" in the bars.

Because the line plot seemed to be an easier representation for students to interpret, there is less that can be said about the individual students. Students T and J again displayed greater competence than students A and G at answering the questions, but this is not surprising given their performance on questions for bar graphs.

STEM-AND-LEAF PLOT TO HISTOGRAM

Stem-and-leaf plots (stem plots) are fairly recent additions to the repertoire of graphical representations. Stems usually represent the 10s digits of the data values and leaves usually represent the 1s digits. This means that the "20s stem" is a row in which data values from 20 to 29 can be placed. However, if 27 is the largest value less than 30, 27 will be the largest value that is actually represented in the 20s stem. This sometimes causes confusion about the values that any stem represents and can make the transition to histogram somewhat confusing.

A histogram represents data that have been grouped in intervals. Stem plots, too, group data in intervals (e.g., by 10s), so it is theoretically easy to imagine rotating a stem plot 90° counterclockwise and visualizing a histogram drawn over the stems. However, in a situation like the one just described, if a learner perceives the data in the 20s stem as extending only from 20 to 27, then the intervals in the histogram might not be drawn (or labeled) with the same width.

In the instructional unit, students were told about the conventions of histograms; that is, the bars touch, each bar represents data in an interval, an interval starts with the lower bound but does not include the upper bound (e.g., the interval 5–10 includes all data from 5 up to and not including 10), and so forth. This information was not, however, the central focus of the unit. The relationship between stem plots and histograms was explored visually by taking one set of data and representing it in stem plots based on 10s and 5s and then turning the stem plots 90° counterclockwise and "filling in" the bars (to cover the leaves) to create histograms.

Interview Responses About Stem Plots and Histograms

The interview questions that we used to explore students' understanding of stem plots and histograms are shown in Fig. 3.9. Because these representations were less familiar to students, it was important to probe their skill at reading information directly from the graphs.

Excerpts from postinstruction interviews are presented in Fig. 3.10. In our study, students in all grades were unfamiliar with stem plots prior to the instructional unit, so preinstruction interviews were not conducted. Further, it was not reasonable to try to assess their understanding of the relationship between the representations. The amount of probing done with the students varied more for this part of the interview than for earlier parts. In part this was due to lack of time toward the end of the interview, and in part it was due to apparently different levels of understanding displayed in students' answers.

After the unit, the eighth-grade students generally seemed able to relate stem plot to the histogram of heights of a fifth-grade class versus heights of basketball players. However, several students "lost" the fifth-grade data value of 171 cm; that is, it sometimes seemed to be ignored and was not incorporated into either the students' or the basketball players' data.

One of the striking features of the students' response was that they could make sense of the two data sets individually, but they had great trouble relating the data sets to each other. For example, they could describe a "typical" student or basketball player, but they did not make the inference that the "typical difference" in heights could be represented by the "difference in typicals." We do not know how students might be best helped in learning to make such inferences, although it seems that discussion of these ideas might be critical for developing students' understanding.

Read the context of the problem situation to the subject.

> *Jane and Sam had some data that showed the heights of professional basketball players in centimeters. They showed it to their classmates. Jane suggested that they make a display that shows the heights of the students in their class and the heights of the basketball players. Then they could answer the question: "Just how much taller are basketball players than students in their class?"*

The class collected and displayed their data using a stem-and-leaf plot or stem plot.

1. Look at the stem-and-leaf plot showing the students in the class. What does the graph tell you about the heights of the students?
   a. How many students are 152 centimeters tall? How can you tell?
   b. How many students are there in the class? How can you tell?

Heights of Students

```
10
11
12
13  8 8 8 9
14  1 2 4 7 7 7
15  0 0 1 1 1 1 2 2 2 2 3 3 5 6 6 7 8
16
17  1
18
```

2. Look at the stem plot showing the basketball players along with the students.
   a. Explain what this graph tells you.
   b. How many basketball players are at least 198 cm tall? How can you tell?
   c. How many basketball players are there? How can you tell?
3. Just how much taller are the basketball players than students in this class?

Heights of Students and Basketball Players

```
10
11
12
13_  8 8 8 9
14   1 2 4 7 7 7
15   0 0 1 1 1 1 2 2 2 2 3 3 5 6 6 7 8
16
17   1
18   0 3 5
19   0 2 5 7 8 8
20   0 0 2 3 5 5 5 5 7
21   0 0 0 5
22   0
23
```

FIG. 3.9. *(Continued)*

4. Here is a histogram of these data. (Show histogram.) How are the histogram and the stem
   plot alike and different? From the data in this graph, how much taller do you think basketball
   players are than students?

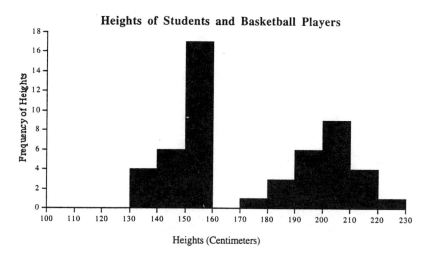

FIG. 3.9. Interview questions to probe stem plot and histogram.

EXTENDING INSTRUCTIONAL TECHNIQUES

The instructional techniques outlined earlier illustrate implicitly one critical
principle that can be generalized to other settings. This is the notion that
establishing connections or translations among representations is critical for
developing understanding. This is not a new idea; for example, Janvier
(1987) and Bowman (1993) made strong cases for the development of translations across representations of functions. Improving understanding of complex mathematical ideas may require focusing the learner's attention on the
translations between representations rather than on the representations themselves (Janvier, 1987). It may not be reasonable to expect most students to
understand the translations if instruction does not focus on those translations.

Based on this principle, it seems that investigations need to be conducted
of how to effectively connect not only other pairs of graphs but also tables
and graphs. For example, box plots might be connected to either line plots
or histograms, and scatter plots might be connected to line plots of the data
represented along each axis separately.

Making correct interpretations of increasingly sophisticated representations
requires more fully developed cognitive functioning. Such representations are
typically built on greater amounts of data reduction, so there are increasingly
many data distributions that might result in identical representations. Correct
interpretations must be worded in a way to account for all of the possibilities

Question	Student T	Student A	Student G	Student J
From the stem plot, what is the typical height of a student?	Anywhere from 147 to 151, . . . cause that's were most of them are.	Between 150 and 158.		
From the stem plot, what is the typical height of a basketball player?	205, because 205 occurs the most than any other.	Between 200 and 207.		
From the stem plot, how much taller are basketball players than students?	About 30 cm shorter. . . . No, about 3 cm shorter . . . cause from 15 to 17 there ain't that—well from 16 to 17 there ain't that many, but then the basketball players start and there's no basketball player shorter than 180 and there's no player, I mean student, really over—the majority of the students are near 150 to 158 and you would skip down to 18 which would be 3 cm or 30 cm . . . 30, that's it.	I don't know.	I would say regular students are really smaller than a basketball player. Basketball players are very big. . . . One student is 17 foot 1 . . . 171 cm. He's almost the size of a basketball player, but everyone else is more smaller—more than 2 feet or 2 cm, like between 16 cm and 17 cm [pointing to the stems of 16 and 17].	Between 10 and 90 cm taller—there's a big range. . . . Here there's only the space of around 10 cm because you go from 171 to 180, and here it would be more like 80 instead of 90 because that's 139 almost 140 to 220 and that gives you about 80 cm.

FIG. 3.10. *(Continued)*

Question	Student T	Student A	Student G	Student J
From the histogram, what is the typical height of a student?	150 to 160.	From 130 to 160.	Around 150 to 160.	
From the histogram, what is the typical height of a basketball player?	From 190 to 210.	From 170 to 230.	200 to 210 cm.	
From the histogram, how much taller are basketball players than students?			There's a big difference in height than students, cause students are really not that big than basketball players. They're really just a little more there to their height, just a little. But they wouldn't be hardly as tall as right here. . . . [The difference in height between a typical student and a typical basketball player] I think it is two, I forgotten how to say it, like two feet difference [pointing to the hole in the data].	

FIG. 3.10. Postinstruction interview excerpts: stem plots to histograms. *Note.* Some cells are empty because lack of time prevented all questions from being asked of all students.

for what the original data might have looked like. That is, interpretations must be general enough to remain true across all of these possibilities.

The instructional unit did not incorporate technology. This was an explicit decision that was driven largely by the lack of technology in the classrooms where the research was conducted. We can imagine that the unit would have been constructed differently if the participating teachers had had access to computers. For example, using graphing programs to make various representations or graphing programs that allowed for multiple, linked representations on the screen simultaneously (e.g., Hancock et al., 1992) would almost certainly have changed not only how data were represented but also what questions would have been asked to focus attention on comparison of representations. Such changes in the instructional unit would have had to have been matched with changes in the ways that students' understandings were assessed.

Instructional Strategies to Help Students Make Connections

One strategy for helping students use tallies as a way to organize data is to ask the students to write their values on "sticky notes" and then to create a line plot using these notes. For example, for the family size problem, each student records the number of children in his or her family on a sticky note; a sample line plot is illustrated in Fig. 3.11.

Students may be questioned about ways to associate the bar (or tower) with corresponding sticky-note values. Students typically are very eager to point out their specific sticky note/bar, even when there are several for the same value. Further questioning can be focused on the labeling of the horizontal axis in the line plot and the relationship of that labeling to the values shown on the sticky notes. That is, the numbers on the horizontal axis show the same information as the numbers on the sticky notes. Students can be asked such questions as:

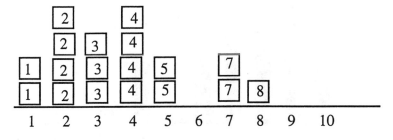

Number of Children in Family

FIG. 3.11. Line plot constructed from labeled "sticky notes."

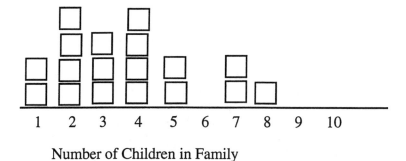

Number of Children in Family

FIG. 3.12. Line plot constructed from unlabeled "sticky notes."

If a sticky note has a 3 on it, what does this mean? (there are 3 children in that person's family)

If a new child joined the class and didn't write anything on her sticky note, but put it above the 6, what does this mean? (there are 6 children in her family)

Display another line plot using unlabeled sticky notes (Fig. 3.12) that shows the numbers of children in 18 different families. What does this graph tell you? How do you know?

These displays use only one axis. To help develop a sense for the use of the vertical axis as a frequency count, students can be helped to translate the "unlabeled display" to tallies as a means of describing the numbers of children in families. Then the tallies can be tied to a vertical axis as a frequency count (Fig. 3.13). This sequencing of activities will not eliminate

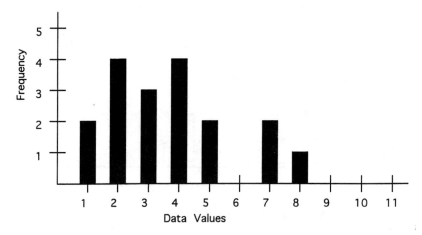

FIG. 3.13. Standard bar graph.

all confusion but does focus on essential components for understanding what it means to "group" data.

The goal of making connections is to help learners internalize ideas about graphs so that these ideas can be used to solve problems. Simply telling learners about the features of graphs will probably not suffice to allow learners to make use of the information. "Generic" pedagogical strategies that might be effective include posing critical questions and asking for learners to reflect on what they are thinking, although more specific strategies related to particular pairs of graphs may ultimately prove to be more important to develop.

CONCLUSIONS

As teachers work with students to help them make translations between representations of ungrouped and grouped data, it seems important for teachers to assure that students are explicitly aware of some of the pitfalls they might encounter. For example, the changing roles of plot elements and axes across representations would seem to be important for learners to recognize. We are certainly not saying the teachers should simply "tell" students about these pitfalls, because we believe that such telling is not likely to be an effective way for students to internalize the important ideas. Rather, teachers probably need to "set up" situations that will allow students to recognize the importance of these changing roles of plot elements and axes.

Further, students need to understand how individual data values are represented in different graphs. Seeing data values explicitly in line plots seemed easier than trying to identify data values implicitly in bar graphs. Also, connecting characteristics of plot elements to other mathematical structures (e.g., knowing how plot elements might be suggestive of models of multiplication) would seem to be important in helping students know how to manipulate information obtained from graphs.

There are some things that teachers might want to avoid doing. Foremost would be assuming that students can make translations between representations easily and quickly. Ideas and observations that seem obvious to teachers are not necessarily obvious to students. Helping students learn which of the critical features of graphical representations need to be attended to will take time. In addition, we suspect that teachers should, at least initially, avoid telling students what procedures to use to solve specific problems. There seems to us to be a strong likelihood that students will overgeneralize these procedures to other representations for which the procedures are not appropriate. Also, students may cease explorations of what representations show if teachers restrict the range of thinking that is identified as acceptable.

We would urge teachers to provide learners with opportunities to compare multiple representations of the same data set. Rich discourse focusing on what each of the representations shows would seem to be necessary to encourage recognition of the similarities and differences in what is communicated by each representation. Probably the biggest challenge for inservice statistics education is figuring out how to help teachers learn to manage this discourse. We need to know much more about how students (and teachers!) think about graphical representations so that critical questions can be formulated that will help learners understand the important features of such representations.

ACKNOWLEDGMENTS

This work was supported in part with funding from the National Center for Research in Mathematical Sciences Education (NCRMSE) at the University of Wisconsin and by the National Science Foundation (grant TPE-9153779). All opinions expressed are those of the authors and do not necessarily reflect the views of NCRMSE, the National Science Foundation, or any other government agency.

REFERENCES

Bowman, A. H. (1993). *Preservice elementary teachers' performance on tasks involving building, interpreting, and using linear mathematical models based on scientific data as a function of data collection activities.* Unpublished doctoral dissertation, University of North Carolina at Greensboro.

Curcio, F. R. (1989). *Developing graph comprehension: Elementary and middle school activities.* Reston, VA: National Council of Teachers of Mathematics.

Friel, S. N., & Bright, G. W. (1995a, October). *Graph knowledge: How students interpret data using graphs.* Paper presented at the annual meeting of International Group for the Psychology of Mathematics Education–North American Chapter, Columbus, OH.

Friel, S. N., & Bright, G. W. (1995b). *Assessing students' understanding of graphs: Instruments and instructional module.* Chapel Hill, NC: University of North Carolina at Chapel Hill, University of North Carolina Mathematics and Science Education Network.

Graham, A. (1987). *Statistical investigations in the secondary school.* Cambridge, England: Cambridge University Press.

Hancock, C., Kaput, J. J., & Goldsmith, L. T. (1992). Authentic inquiry with data: Critical barriers to classroom implementation. *Educational Psychologist, 27,* 337–364.

Janvier, C. (1987). Translation processes in mathematics education. In C. Janvier (Ed.), *Problems of representation in the teaching and learning of mathematics* (pp. 67–71). Hillsdale, NJ: Lawrence Erlbaum Associates.

Joram, E., Resnick, L. B., & Gabriele, A. J. (1995). Numeracy as cultural practice: An examination of numbers in magazines for children, teenagers, and adults. *Journal for Research in Mathematics Education, 26,* 346–361.

Kader, G., & Perry, M. (1994). Learning statistics with technology. *Mathematics Teaching in Middle Schools, 1,* 130–136.

National Council of Teachers of Mathematics. (1989). *Curriculum and evaluation standards for school mathematics.* Reston, VA: Author.

CHAPTER FOUR

Teach-Stat: A Model for Professional Development in Data Analysis and Statistics for Teachers K–6

Susan N. Friel
University of North Carolina at Chapel Hill

George W. Bright
University of North Carolina at Greensboro

The *Curriculum and Evaluation Standards for School Mathematics* (National Council of Teachers of Mathematics [NCTM], 1989) identifies statistics and probability as a major strand across all grade levels. Since 1989, there has been growing interest in what to teach and how to teach with respect to statistics. It is only in the past few years that appropriate curricula (e.g., Used Numbers Series,[1] Quantitative Literacy Series[2]) have become available for use at the K–12 levels.

At the elementary level, available curricula are an essential ingredient in helping teachers find ways to integrate the teaching of statistics in a coherent and comprehensive manner. However, curricula for use with students are not sufficient; using such curricula effectively requires a reasonable knowledge of statistics in order to pose tasks appropriately and promote and manage classroom discourse successfully. In many cases, the only exposure elementary teachers may have had to statistics was at some point during their teacher preparation programs in learning to read and interpret research. Usually, this experience is not remembered as a positive one; for many, what was learned then is not applicable to creating an instructional environment now to help students learn statistics and learn to use the process

---

[1] *Used Numbers: Real Data in the Classroom*, a set of six units of study for K–6 students, is published by Dale Seymour Publications (Palo Alto, CA).

[2] *Quantitative Literacy* Series, a set of four units of study for Grades 8–12 students, is published by Dale Seymour Publications (Palo Alto, CA).

of statistical investigation. Elementary teachers are in need of professional development opportunities that will support their learning of content and promote the use of an inquiry orientation to help their students learn and use statistical concepts.

Teach-Stat: A Key to Better Mathematics was a project[3] designed to plan and implement a program of professional development for elementary teachers, grades 1–6, to help them learn more about statistics and integrate teaching about and teaching with statistics in their instruction. This project included three components:

1. The design of professional development curricula for use with teachers and with teacher leaders (here referenced as statistics educators)
2. A large-scale implementation program to provide professional development for both teachers and statistics educators using the professional development curricula
3. A program of research and evaluation to assess the impact of the project and to surface research questions related to the agenda of the project

PROJECT DESIGN

This project was funded by the National Science Foundation through the University of North Carolina Mathematics and Science Education Network (MSEN). MSEN consists of 10 centers across North Carolina housed at 10 of the state university system's campuses. Each center is directed by a faculty member in mathematics or science education; one of the main tasks for each center is providing professional development in mathematics and/or science for teachers K–12 in its service region. Because of its structure, MSEN is particularly well suited for supporting the implementation of large-scale, statewide projects.

The project involved 9 of the 10 MSEN centers; one faculty member (here referenced as site faculty leader) from each of the sites served as the local coordinator of the project. The 9 site faculty leaders, with the addition of a few other university consultants, designed and implemented the Teach-Stat project. Over 450 teachers across North Carolina participated in the project and, of those, 84 Teach-Stat teachers received additional professional development to prepare them to be statistics educators.

The first fall and spring (1991–1992) of the Teach-Stat project was used as a planning time to bring together the site faculty leaders as well as

[3]Friel, S. N., & Joyner, J. M., Principal Investigators. *Project Teach-Stat: A Key to Better Mathematics.* NSF grant TPE-9153779.

additional faculty consultants from across the state. This group met intensively for two- to three-day meetings several times during the first year. Their tasks were to design the draft of the professional development curriculum that would be used to teach teachers and then to jointly teach the first 3-week summer institute (note: *institute* and *workshop* are used interchangeably).

The project was designed so that, during the first year, each site faculty leader selected 6 or 7 teachers as a pilot team. The 57 teachers and 9 site faculty leaders participated in the 1992 3-week summer institute, which was offered as a residential program at a central site. The faculty, working in teams of three, was responsible for various parts of the program. Throughout the following school year, each faculty site leader met with and visited the regional teacher team, jointly exploring with the teachers what it meant for them to teach statistics and integrate statistics with other subject areas.

In the second year, each of the 9 sites offered a revised (nonresidential) version of the 3-week professional development program to 24 new teachers. Each site faculty leader and the pilot team of 6 or 7 teachers worked together to plan and deliver the workshop. Originally, the pilot teachers were going to be available to help but not to teach. However, by the second summer, faculty and teachers had developed such a good working relationship that the model of a "professional development team" naturally emerged and was very successful. The second-year participants were able to hear from teachers who had spent the preceding year teaching statistics and had lots of actual examples to show them. The first-year pilot teachers received a great deal of support, informal "how to be a staff developer" training, and coaching/mentoring from their respective site faculty leaders.

The site faculty leaders provided the support for Teach-Stat teachers following the first- and second-year institutes. Following the first-year insitute, site faculty leaders generally made at least two classroom visits to each teacher and often included visits to principals as well. In addition, site faculty leaders and teachers met together as a team on several occasions. Following the second-year institute, the number of Teach-Stat teachers at each site increased to about 30 teachers so that site visits were more limited in scope.

In the third year, 84 teachers from either the first or second year were selected to become statistics educators in order to serve as resource people across North Carolina to provide professional development programs in statistics education for other elementary teachers. They participated (regionally at the nine sites) in the equivalent of a 1-week seminar focused on the "how-to's and why's" of staff development. The statistics educators at each site were responsible for developing and delivering a 2-week summer institute for an additional set of 24 new teachers at their site. As a result of this program, the statistics educators were equipped to offer the Teach-Stat professional development program to other elementary teachers as well as to design variations of this program to meet the needs of the audience with which they happened to be dealing.

The documentation of the project includes various materials that permit others to replicate the program of professional development and implementation.

1. *Teach-Stat for Teachers: Professional Development Manual* (Friel & Joyner, 1997). Provides a "how-to" discussion for planning and implementing a 3-week teacher education institute. Its audience is mainly those who provide professional development programs for elementary teachers.

2. *Teach-Stat for Statistics Educators: Staff Developers Manual* (Gleason et al., 1996). Provides a "how-to" discussion for planning and implementing a 1-week Statistics Educators Institute. It is designed for teachers who will serve as staff development resource people (statistics educators), have participated in a 3-week program in statistics education, and have previously taught statistics to students.

3. *Teach-Stat Activities: Statistics Investigations for Grades 1–3* and *for Grades 3–6* (Joyner et al., 1997a, 1997b). Provides activities for elementary students, many of which were used by Teach-Stat teachers, that promote the learning of statistics using the process of statistical investigation.

DESIGNING PROFESSIONAL DEVELOPMENT CURRICULA

Teach-Stat for Teachers—Professional Development Program

In a recent article, Gravemeijer (1994) articulated an alternative model for the well-known "research–development–diffusion" model of curriculum development. He characterized a process of curriculum development as one of "educational development" that is based on an integration of curriculum research and design and can be described as a process that not only is guided by theory but also produces theory.

> In general, curricula are developed to change education, to introduce new content or new goals, or to teach the existing curriculum according to new insights. (p. 445)
>
> Curriculum development can very well be seen as constituting a composition of instructional activities that makes sense to the developer. Making sense in this case means that the set of instructional activities has the character of a theory on how to ensure that the students learn what is intended. (p. 448)

Gravemeijer drew distinctions between curriculum development and developmental research. In the former, the idea is one of using what is available and adapting those means to one's momentary goals. In the latter, the evolutionary aspect is much more important in the sense of a goal-oriented process of improvement and adjustment; that is, the process is guided by theory that grows during the process. How to develop instruction is explored as a part of the process of educational development (pp. 443–451).

Gravemeijer's views may be applied to an emerging area, that of the design of professional development curricula used to support teachers in making changes in their teaching. Indeed, a central component of the Teach-Stat project was the articulation of what is known (or came to be known during the 3 years of the project) about theory and practice as it relates to the teaching and learning of statistics for K–6 teachers. In designing a set of instructional activities that made sense in this situation, the developers[4] were guided (a) by beliefs[5] about what statistics is, how it is learned, and how it should be taught both to teachers and to students and (b) by a variety of shared experiences including extensive opportunities for planning and a first-year summer (residential) institute that was jointly team taught by all faculty.

The process of statistical investigation became the focus for the instructional model developed during the project. The developers spent a number of sessions articulating their beliefs and then framing a coherent curriculum that supported teachers' learning of statistics in an environment that both modeled and encouraged teachers' eventual use of the key components of teaching as articulated by the *Professional Standards for Teaching Mathematics* (NCTM, 1991). To do this, the developers worked to "get past" the notion of putting together a set of activities that addressed selected statistical concepts, because developing a list of activities did not address the process of teaching and learning that was believed to be central to the program.

Two theoretical perspectives helped shape this direction. One was the conception of statistics as a process of statistical investigation and the articulation of this process by Graham (1987) and Kader and Perry (1994). The other was the introduction of use of concept maps (Novak, 1984) during the first-year summer institute as a way of assessing what teachers knew about statistics prior to and following the institute. Following this institute and during

---

[4]As part of a year-long planning process for planning the first institute, all faculty read Moore's article, "Uncertainty" (1990), explored the Used Numbers program, and were updated on current research related to statistics teaching and learning. The makeup of the faculty was one science educator, three statisticians, eight mathematics educators or mathematics educators/mathematicians. This composition provided a diversity of perspectives, and contributions highlighted the different strengths each faculty member brought to the project.

[5]A very direct influence, in addition to those noted earlier, was the NCTM *Professional Standards for Teaching Mathematics* (1991). The four components of teaching—task, discourse, environment, and analysis—were central to the instructional model developed for use in the Teach-Stat project.

a revision phase of the first draft of the professional development curriculum, a project concept map was developed as way to show the key ideas the developers sought to address in the Teach-Stat materials. The concept map (Fig. 4.1; Friel & Joyner, 1997) that has evolved from the Teach-Stat project includes the four components of the statistical investigation process and elements of statistics content that relate to these components. The center of the map displays the four components of the process of statistical investigation. Attached to each part of the process are related statistical concepts. This process and its related concepts may be described as follows:

- When we pose the question, we do so because we want to solve a problem. The problem involves exploring one or more directions: describing a set of data, summarizing what we know about a set of data, comparing and/or contrasting two or more sets of data, or generalizing from a set of data in order to make predictions about the next case or about the population as a whole. The specific question asked is related to the problem being explored and determines what data need to be collected. The types of data (e.g., categorical or numerical) dictate the methods for data collection (e.g., counting or measuring).

- Collecting the data involves identifying the actual population to be studied and the methods for data collection. If sampling is involved, different types of samples may be considered, including random samples, convenience samples, or a census. This is the point at which bias, representativeness, and randomness as they relate to sample selection become important.

- The analysis of the data may include describing and/or summarizing a single set of data, comparing and/or contrasting two or more sets of data, or making predictions and/or assessing implications from one or more sets of data . Methods for doing this involve organizing, sorting, classifying, and displaying data using tables, diagrams, and graphs, and determining descriptive statistics, such as measures of central tendency (mean, median, and mode), measures of variation (range and standard deviation), and measures of association (line of best fit and correlation coefficients).

- Interpreting the results leads us back to the original question posed and to the purpose(s) for the investigation. How do the data we have collected and analyzed help us respond to the specific question posed?

The concept map displays statistical concepts that help define what teachers and students need to know about statistics. The process of statistical investigation defines what they need to be able to do.

The emphasis on viewing statistics as a process of investigation and then framing a concept map that helps articulate the relationship among the process and key concepts evolved during the first 15 months of the project.

Concept Map of the Process of Statistical Investigation

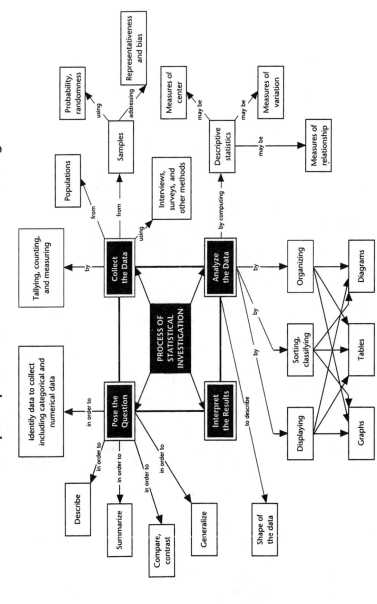

FIG. 4.1. Concept map of the process of statistical investigation. From *Teach-Stat for Teachers*. © 1997 by The University of North Carolina by Susan N. Friel and Jeane M. Joyner (Eds.). Published by Dale Seymour Publications. Reprinted with permission.

The emerging changes in the curriculum were based on a deeper understanding of statistics content and learning and reflected the kind of tinkering
between curriculum and theory that Gravemeijer discussed.

The materials developed for *Teach-Stat for Teachers: Professional Development Manual* (Friel & Joyner, 1997) are framed with this concept map in
mind. The explorations of statistical concepts are not conducted in a specified
order related to the flow of the concept map (i.e., pose the question, collect
the data, analyze the data, interpret the data). Rather, each activity is structured to be a statistical investigation with the four components of the process
used to define the development of the activity. The various activities provide
special emphasis on the components of the process and the related statistical
concepts.

As an example of how the four phases of a statistical investigation may be
woven together to provide a format for activities, the first activity, called About
Us, is shown in Fig. 4.2. The purpose of this activity is to focus participants'
attention on the process of statistical investigation and also to clarify what
knowledge and skills they bring to the instructional setting. Notice that the
activity itself is framed as a statistical investigation using the headings of pose
the question, collect the data, analyze the data, and interpret the data.

If statistics is to be considered a topic in the elementary grades, there is
a need to develop ways to help students and their teachers visualize and
experience concepts in meaningful ways. As one example, the concept of
measures of center—mode, median, and mean—is an important part of the
analysis of the data component in the process of statistical investigation.
The mode is the value that occurs most frequently in a data set. It may be
visualized on a line plot or a bar graph as the "highest tower" on the graph.
The median is the middle value in an ordered set of data. It may be visualized
by listing the data values in order, one value per square on a strip of square
grid paper. The strip is folded in half to locate the median, with the fold
landing exactly on a single value (in the case of an odd number of data
values) or between the two middle data values (in the case of an even
number of data values).

With respect to the mean, research (Mokros & Russell, 1995) has suggested
that it is a difficult concept for students to grasp. Not only may it be better
to delay its introduction to upper elementary grades (grades 5–6), it also is
helpful if teachers (and then their students) have concrete images that allow
them to visualize what the mean is.

Project staff and site faculty leaders explored two different models for
visualizing the mean: the balance model (Friel et al., 1991) and the evening
out model (see Fig. 4.3, Investigation 26, as one of a series of activities
focused on the conceptual development of the mean). The goal for this
work with teachers and then their subsequent work with their students is
to have the concrete model provide a justification for the development of

About Us

Overview

Teams of three to six participants each frame a question they would like to investigate about the entire group of participants. This question becomes the basis for a mini-research project that teams complete during the workshop. Participants use the process of statistical investigation to complete an initial investigation of their questions, reporting preliminary findings to the whole group.

Assumptions

Participants have different backgrounds with respect to their knowledge about and experience with statistics and the process of statistical investigation. While they may be uncomfortable with the open-endedness and apparent lack of structure of the activity, engaging in an activity that requires them to develop their own structure encourages them to impose an understanding of their current knowledge in this area. This process of inquiry promotes discourse and the development of a mathematical community.

Goals

Participants explore the process of statistical investigation. In particular, they

- develop an overview of the process of statistical investigation: posing a question, collecting the data, analyzing the data, and interpreting the results
- recognize the need for refining or redefining the problem situation as part of the process of statistical investigation
- demonstrate current knowledge of key statistical concepts: representations, measures of center, measures of variation, and sampling

Teacher Notes

This is always enjoyed as a beginning learning activity.

Because participants may at first experience a high degree of anxiety, it may be helpful to choose groups carefully. Allow people to be with whom they are most comfortable.

Materials

Pencils, pens, graph paper, lined paper, scissors, colored markers, calculators, colored stick-on dots, chart paper

Transparencies 1.a, 1.b

Handout 1.1

Reference

Rosebery, A. S. B. Warren, R. Corwin, A. Rubin, F. Harik, F. R. Conant, and S. Friel. *Reasoning About Data: A Course in Statistical Inquiry for Teachers of Middle School, Grades 5–9*, Cambridge, Massachusetts: Bolt Beranek and Newman, 1990.

Developing the Activity

The development of the activity encompasses the four interrelated components of the process of statistical investigation.

Pose the Question

In small teams of three to six people, participants consider the following problem situation as it pertains to all members of the workshop:

What would you like to know about yourselves as a group?

Each team selects a question they would like to investigate about the group (for example, education level and major, amount of reading done each week, or distance traveled to the workshop).

Sample Questions

What kind of coffee do you drink?

What are your hobbies?

How much television do you watch?

How much do you read? What do you read?

How do you get to work?

If you weren't an educator, what field would you be in?

Where were you born?

Collect the Data

After identifying their questions, each team develops or invents their own methods for collecting the data. Depending on the size of the group, a variety of issues may surface.

- Did you ask the same person I did?
- If the group is much larger than 30 people, do we have to ask all of the people our question?
- Our question raises lots of questions about definitions—what do we mean by "How much reading do you do?"

Encourage discussion, but let each team make decisions about their actions without access to the advice of experts in the field.

Teacher Notes

Teachers may want to informally check among themselves that each team is investigating a different question, although investigating the same question can be profitable, tool

Analyze the Data

To report their findings, each team should represent their data graphically. They may make use of such strategies as numerical representations, measures of center, measures of spread, symmetry,

FIG. 4.2. *(Continued)*

and observations about the shape of the distribution (clusters, gaps, unusual values) and patterns that emerge. Discussion among team members as they analyze their data is invaluable in highlighting the process of statistical investigation and the components involved.

Interpret the Results

Teams may draw conclusions and raise a number of questions in the process. It is likely that each team will have additional ideas for carrying out their investigations in ways other than those first suggested by their initial questions. This is an important part of the process of statistical investigation and, in this instance, forms the foundation for continuing this activity as a mini–research project to which teams return later in the workshop.

Summary

Each team presents their findings to the group of participants. As each team presents its results, encourage the team and the group to articulate issues, questions, and observations about the statistical investigation process. Note their ideas on the board.

Following the presentations, review the issues and observations that were raised and group them according to the corresponding process of statistical investigation. Use Transparency 1.a to introduce the four components, letting participants decide where each question or issue belongs.

Teacher Notes

It's a good idea to post these graphs and representations to return to throughout the workshop.

- *Pose the problem.* Are problems framed so that it is clear what information is being sought? What is the purpose for collecting the data: to describe, to summarize, to compare, or to generalize by predicting information about the next case or the population?

- *Collect the data.* How were the data collected?

- *Analyze the data.* What different ways were used to report the data (for example, table, line plot, bar graph, summary statistics)?

- *Interpret the data.* Return to the original purpose for data collection: to describe, to summarize, to compare, or to generalize. Discuss representativeness of data, including sample size. What other questions come to mind based on these data?

Use Transparency 1.a and 1.b to clarify concepts related to each component of the process. Distribute Handout 1.1.

Teacher Notes

Samples of the kinds of issues that might surface

People didn't know what was meant by family so we got different data from people.

Was any person sampled twice?

Was everyone asked?

Are the data biased?

Are there other ways to report the data?

Is our group representative of a larger population? How can we tell?

Is there a relationship between type of coffee consumed and gender?

About Us Project

As a follow-up, teams reconsider their question(s)—they may even change the question(s)—and complete the process again, preparing a final presentation for the last day of the workshop. Note that in the three-week syllabus, there is time allocated for this as well as for meeting in teams the day before presentations to complete planning.

FIG. 4.2. Activity incorporating the statistical investigation process. From *Teach-Stat for Teachers.* © 1997 by The University of North Carolina by Susan N. Friel and Jeane M. Joyner (Eds.). Published by Dale Seymour Publications. Reprinted with permission.

INVESTIGATION 26
Family Size Revisited

Overview

Participants explore building a model for the concept of the mean. They begin by "evening out" data values to find the mean. This technique is linked to "balancing a distribution," the idea that any deviation from the mean in one direction must be matched, or balanced, by an equal deviation in the opposite direction.

Assumptions

Participants have previously considered and collected data for the problem of the typical family size in Investigation 13, *Family Size*. They know how to compute the mean and median of a data set.

Goals

Participants gain an understanding of the mean as a measure of center. In particular, they

- describe data by estimating the mean

- construct sets of data for a given mean

- understand the mean as a way of evening out a set of data and as a balance point in a set of data

- understand that the mean alone may not give enough information about a set of data

References

Friel, S. N., J. R. Mokros, and S. J. Russell. *Statistics: Middles, Means, and In-Betweens*. Palo Alto, California: Dale Seymour Publications, 1992.

Fey, J., W. Fitzgerald, S. Friel, G. Lappan, and E. Phillips. *Data About Us*. Palo Alto, California: Dale Seymour Publications, Forthcoming.

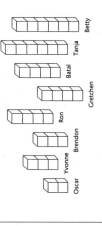

Materials

Stick-on notes, graph paper, colored markers, interlocking cubes in 8 colors, calculators

Transparencies 26.a, 26.b

Handouts 26.1, 26.2

Developing the Activity

Participants will explore representations of the mean using towers made of interlocking cubes and line plots made of stick-on notes.

Pose the Question

Remind participants of their work in Investigation 13, *Family Size*. Ask them how they defined a family in that investigation.

Discuss examples of averages from school, the community, the news, and the popular media.

Remember that an average is a single number or value often used to describe what is typical about a set of data. It can be thought of as a "measure of location." The median is one kind of average that we have used quite a bit. It shows us the location at which a set of ordered data is divided in half: half the data are below that point, and half the data are above it. Now you will explore another kind of average, the mean.

Pose this problem for participants to consider.

Eight students in one class gathered data about the number of people in their families. Each student made a tower of cubes to show the number of people in his or her family. Here is what they found.

As you describe the problem situation, display eight towers of interlocking cubes. Use a different color for each tower. The first letter of each name suggests a color to use (such as Yvonne for yellow; Brendon, Batal, and Beth for black, dark blue, and light blue).

FIG. 4.3. *(Continued)*

Work with participants to see the frequencies of data values implicit in the cube towers and then make a line plot with stick-on notes to show the data.

Number of People in Students' Families

Number of people

Discuss how many people there are in the eight families. Make sure participants can obtain this information from both the cube towers and the line plot. Explain that students may not immediately see that the cube towers and the line plot are representations of the same information.

Analyze the Data

Ordering the data towers enables us to group by number of people.

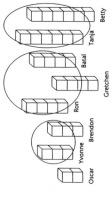

Oscar Yvonne Brendon Ron Gretchen Batai Tanja Betty

The students wondered how many people there are in the eight families on average. What is the mode family size for these eight families? How is it shown with the cubes? How is it shown with the stick-on note line plot? Do they show the same mode? Why might students be confused with what the mode is?

Students often think of the mode as the *most* in a data set and miss the subtle distinction needed to understand that the mode is the *most frequent* data value. It is not unusual for a student to say something such as, "The mode in the towers is 6 and the mode in the line plot is 4." This reflects students' lack of connection between the two representations.

What is the median family size for these eight families? How can you find it using the cube towers? How else can it be found?

The strategy students will most likely use with the cube towers is to work in from the ends: pairing the shortest and tallest towers, removing each pair until they reach the middle two data values, then computing the median.

The students found another way to find an average: by moving people to other families to "even out" the sizes of the families. They know they must keep a total of 32 people in the eight families.

With participants, add the cubes that are in all the towers.

Oscar	2 people
Yvonne	3 people
Brendon	3 people
Ron	4 people
Batai	4 people
Gretchen	4 people
Tanja	6 people
Betty	6 people
Total	32 people

The students moved cubes from one tower to another, making some families bigger than they actually were and other families smaller than they actually were. When they were finished, their cube towers looked like this.

Move cubes from taller towers to shorter towers until every tower has four cubes.

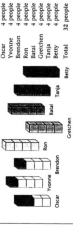

Oscar	4 people
Yvonne	4 people
Brendon	4 people
Ron	4 people
Batai	4 people
Gretchen	4 people
Tanja	4 people
Betty	4 people
Total	32 people

Because you used different colors for each family, participants will be able to see how the evening out is accomplished. You may want to highlight the number of people in each family as shown both before and after the evening-out process, and the total number of people in all the families.

FIG. 4.3. (Continued)

This representation shows another kind of middle number. We can say that, on average, there are 4 people in each family. We call this number the mean. The mean number of people in the eight families is 4. There are a total of 32 people in the eight families.

Display Transparency 26.a, which shows the line plot the students made for the eight families and depicts the mean as a balance point. Distribute Handout 26.1, and discuss the questions posed on it.

Notice that the mean is shown as a location on the number line and is indicated by means of an arrow. The mean is a kind of balance point. We can show that many family sizes are not the same as the mean but are related to the mean. Some are less than the mean, and some are more than the mean.

Another group of eight students in the class had the following data about the number of people in their families.

Again, make eight cube towers, using distinctive colors for each. Then make a line plot to show this new set of data.

Radford Tishia Bert Gabi Byron Yuri Orrie Bonnie

Number of People in Students' Families

Number of people

What is the mean number of people in these eight families? How can we show a balance point for the data displayed on the graph?

Display Transparency 26.b. Again, point out that the sum of the differences from the mean for data values below the mean is the same as the sum of the differences from the mean for data values above the mean.

Have participants work in small groups to consider the questions on Handout 26.2.

Interpret the Results

It is possible to create other sets of eight families with a mean of 4 people. Working with a partner, see if you can find two more sets of eight families with a mean of 4 people. Use cubes to show each set, then make line plots that show the information from the cubes.

Distribute Handout 26.3, which will give participants more experience with these concepts.

Summary

This is the first part of our introduction to the mean. The most important outcome is flexibility in moving between the representations of the data and being able to talk about the mean and how it can be pictured.

FIG. 4.3. Introducing the balance model for the mean. From *Teach-Stat for Teachers*. © 1997 by The University of North Carolina by Susan N. Friel and Jeane M. Joyner (Eds.). Published by Dale Seymour Publications. Reprinted with permission.

the algorithm. After a number of experiences with children and with teachers, the developers settled on the evening out model as a more effective choice. The balance model lends itself well to looking at variation within the data and has potential when considering such measures as absolute mean deviation or standard deviation.

The developers spent a great deal of time making decisions about the role of probability as part of the institute. A limited set of activities are included that focus on probability as it relates to using such manipulatives as tossing dice or coins; the approach to these investigations is conducted within the framework of the process of statistical investigation. These experiences are used to launch the notion that dice can be used to help randomly select samples of subjects. Two 10-sided dice and a database of 100 cats with relevant statistics such as weight, length, color, age, and so on are used to demonstrate the concept of random sampling. Groups of teachers select samples of 20 (without replacement) from this database and compare their data with the population data, with data collected from such samples as all females or from the population of cats, and with other groups' data. This development provides teachers with a concrete understanding of what it means to select a random sample from a population as well as how a sample relates to the population from which it is drawn.

Throughout the program, issues of representativeness and bias are discussed as they relate to the samples being used and teachers' potential for drawing conclusions and making predictions about some larger population. Much of the data collection that is done involves use of convenience sampling and/or self-selected samples. This is intentional as participants (and their students) need to be actively involved in collecting their own data from sources that are both convenient and available. For the most part, there is very little reason to do otherwise; the focus is on descriptive statistics and the process of statistical investigation. However, the work done with probability building the linkage to random sampling gives participants a beginning understanding of what random sampling is and why it is necessary.

Teach-Stat for Statistics Educators—Staff Development Program

A similar design process was used to prepare the program that was used to help the 84 statistics educators explore what it means to provide professional development for teachers. This component of the project focused on providing staff developers, such as university faculty and school administrators, with plans for helping teachers who are already implementing Teach-Stat in their classrooms explore ways to work with other elementary teachers so they can learn to teach statistics.

The format of the Statistics Educator Institute provides opportunities for statistics educators to plan their first Teach-Stat workshop. A central component of the program is supervised teaching of statistics educators during a regular 2- or 3-week Teach-Stat workshop. The staff developer (in the case of the project, the site faculty leader) facilitates the planning and serves as a resource and coach when the statistics educators present a Teach-Stat workshop.

The development of the curriculum was a dynamic process, involving the nine site faculty and their teams of statistics educators. After the second year of the project, faculty met to plan the Statistics Educators Institute. A draft document detailing the program was in place by spring. Because of the nature of the teachers who were involved, that is, had participated in Teach-Stat earlier, with many already working with small groups of teachers in their school districts, the actual implementation of the planned program involved joint interaction among the site faculty and their teams. Indeed, very possibly each team "experienced the program" differently simply because of the interaction of past experiences with Teach-Stat and with their respective site faculty as well as other factors. These interactions provided rich opportunities for modifying and redesigning the statistics educators manual.

The observations and work with the teachers during the first and second years of the project provided some of the material for the curriculum. For example, after the first-year pilot teachers had completed working with their site faculty leaders in the second-year workshops, a number of these teachers were interviewed. Subsequently, a select number of these interviews were incorporated in the *Teach-Stat for Statistics Educators* curriculum as part of an activity called "Teacher Talk" (Fig. 4.4) to be used as cases to be analyzed in light of what is known about teaching from the perspective of the *Professional Standards for Teaching Mathematics* (NCTM, 1991). Statistics educators were asked to analyze each case to discuss whether the teacher had changed in the directions suggested in the teaching standards.

In another section in this curriculum, statistics educators revisit issues of content. Even after participating in a 3-week institute that focused on content and after teaching their own students for at least a year with the support of a visiting site faculty leader, many teachers still had "holes" in their understanding of statistics content. When site faculty leaders visited teachers, some of the confusions about how to handle content were observed as teachers worked with their students. In response to these content concerns, several vignettes were created to highlight specific content issues (Fig. 4.5). Each vignette presents a "content dilemma"; statistics educators respond to one or more questions about the situation. The "content dilemmas" selected as vignettes were chosen because site faculty leaders observed variations of these situations during classroom visits to Teach-Stat teachers in the earlier part of the project.

Teacher Talk: Rachael

We did some things with pumpkins in the fall. We were interested in the circumferences of pumpkins. And we did the circumference and talked about measuring. And then our question was, "Do we think that, based on the circumference of the pumpkin, the bigger ones would have more seeds?" And of course the kids were sure that the big ones had to have more seeds. So we actually counted. We cut open several pumpkins. Once we did that, we could see that the bigger pumpkins did not necessarily have more seeds than the smaller ones. And we talked about that quite a bit, and we graphed that. And a neat idea that came out: one girl said she had a belt of chain links and wanted to measure the pumpkins with chain links. They discovered that one of the pumpkins was as big around as the girl's waist. And we talked about how big around things are.

The positive aspects of being in Teach-Stat are understanding mathematical principles more and understanding how statistics fits into all areas of the curriculum and all that we do. We can look at statistics and some type of categorization of the data, and we can make it make more sense. Statistics really is the key to helping understand our world even better. And we can look at things we do in the classroom and base them on what they're doing out in the world and tie the world in through the vehicle of statistics. And I didn't understand that before. I didn't really think about that before we had Teach-Stat.

FIG. 4.4. Teacher Talk activity. From *Teach-Stat for Statistics Educators.* © 1996 by The University of North Carolina by Jane M. Gleason, Elizabeth M. Vesiland, Susan N. Friel, and Jeane M. Joyner (Eds.). Published by Dale Seymour Publications. Reprinted with permission.

Again, the curriculum designed for working with statistics educators reflects the kind of tinkering described by Gravemeijer. An evolving process of improvement and adjustment took place, leading to the arbitrary completion determined by the end of the project itself. With publication of the materials, it is hoped that others will use these materials not as an end in themselves, but rather, in the sense of Gravemeijer's discussion of developmental research, as an evolving process of improvement and adjustment.

Vignette: Median

Overview

This vignette illustrates paths that students may take as they construct their conceptions of median. How many different conceptions and misconceptions of median can you find in the dialogue?

Vignette

When Mrs. English's class investigated the typical number of raisins in a half-ounce box, the following data were recorded:

40, 35, 35, 30, 30, 28, 31, 29, 35, 36, 28, 29, 28, 38, 38, 29, 31, 31, 28, 28, 30, 30, 32, 34, 35, 35, 38, 34

Mrs. English: You may recall that we began talking about the median of a set of data yesterday. Sometimes looking for the median helps us decide what is typical. Look at these data. Think for a few minutes about what you think would be the median of these data.

Shawn: It's 15. There are 30 numbers, and half of 30 is 15. So 15 is the middle, or median.

Paulina: No, the median is 38, because there are 30 boxes of raisins and half of 30 is 15, and the fifteenth number is 38.

Sook Leng: I disagree. First you have to put your numbers in order from smallest to largest like this:

28, 28, 28, 28, 28, 29, 29, 29, 30, 30, 30, 31, 31, 31, 32, 34, 34, 35, 35, 35, 35, 36, 38, 38, 38, 40

Now 31 is the median because there are 30 boxes of raisins and half of 30 is 15 and the fifteenth number is 31.

Erin: I almost agree with Sook Leng, but the median has to have the same number of boxes on each side. So, you have to say that the median is between 15 and 16, and that would be 15½.

Jo Anne: I've been thinking about what Sook Leng said. You could think of 31 as the median in another way. I made a line plot. If you look at the tallest towers, they're at 28 and 35. Then you can find the middle of these two towers. That would be 28 + 35 = 63, and 63 divided in half is about 31. So 31 must be the median.

Jo Anne's Line Plot of the Class Raisin Data

```
                                    x
x                                   x
x           x                       x
x           x           x           x       x
x       x   x   x   x   x   x   x   x       x
x       x   x   x   x   x   x   x   x   x   x
26  27  28  29  30  31  32  33  34  35  36  37  38  39  40
              Number of raisins
```

Mrs. English: Joseph, do you agree that 31 is the median of these data?

Joseph: No, I'm thinking the median is 34. See, the smallest number of raisins found was 28, and the largest was 40. If you list these numbers in order, like 28, 29, 30, 31, 32, 33, 34, 35, 36, 37, 38, 39, 40, the middle number is 34, because there are 13 numbers, and you need to find the number with six numbers on each side of it, like six numbers above it and six numbers below it.

Michael: I have another idea. The median is 33, because the numbers at the bottom go from 26 to 40. They would balance at 33. There would be seven numbers above 33 and seven numbers below.

Mrs. English: I'm not sure about that, Michael. Is 33 the median of the data, or is 33 the median of the scale?

Michael: Oh, I see. So 33 would be the median of the scale. But 33 can't be the median of the data, because there aren't any data there! So I guess there isn't a median, or it might be zero since there isn't an X there.

Carl: I'm not so sure. I think the median of these data is 32, because when you look at the plot you see that most of the boxes had between 28 and 36 raisins, and the middle of 28 and 36 is 32.

Brennan: I agree with Michael. The median is 33, but it's because there are 10 towers of Xs. At 33 we'd have five towers on each side.

Discussion Question: Median

What conceptions and misconceptions about median are in this vignette?

FIG. 4.5. Vignette highlighting content issues. From *Teach-Stat for Statistics Educators.* © 1996 by The University of North Carolina by Jane M. Gleason, Elizabeth M. Vesiland, Susan N. Friel, and Jeane M. Joyner (Eds.). Published by Dale Seymour Publications. Reprinted with permission.

PROJECT FINDINGS AND OUTCOMES

Content and Pedagogy Surveys

We used a variety of assessment and evaluation strategies throughout the project. In order to evaluate changes in content and pedagogy, we developed two written instruments that were then used extensively with first-year participants (and less extensively with second-year participants) to describe changes they experienced over time.

The Statistics Content Survey was developed prior to the design of the professional development curriculum and contains a number of open-ended questions that related to many of the concepts that were addressed as part of the professional development program. The results suggest that Teach-Stat was very successful in increasing teachers' understanding of statistical concepts (Gregorio et al., 1995). The greatest change in content knowledge was noted between the pre- and postinstitute assessments. We believe that teachers' knowledge about statistics continued to change in subtle ways after the institute, but our content instrument failed to provide us with needed insights. A few select and quick visits to statistics educators' classrooms have raised project staff's awareness that much may be ascertained about teachers' understandings of content by watching them teach. However, this was not a strategy that was systematically implemented in this project.

The results from the several administrations of the content survey, however, did influence the professional development curriculum. After the summer institute, we discovered that teachers continued to be quite confused about median as a measure of center. This encouraged us to revise our investigations to support a deeper understanding of both how to identify the median in a set of data and how to use the median as tool for describing and/or comparing sets of data.

As another example, the developers found that teachers demonstrated some confusions about reading data from a bar graph. One general confusion that emerged was that some teachers misread the axes, looking for data values instead of frequencies on the y-axis. This observation supported informal experiences that statisticians had noted with their college students. It also supported informal observations made through work one of the principal investigators had with children and teachers in an earlier project, that is, the Used Numbers project. Because of this problem, a few of the developers have since designed and carried out a research project looking at how students read graphs (see Bright & Friel, chapter 3, this volume). Further, an additional mini-lecture[6] was added to the Teach-Stat professional

---

[6]Throughout the curriculum found in *Teach-Stat for Teachers: Professional Development Manual* are a number of mini-lectures. The purpose is to summarize what is known (primarily from research) about how students learn key concepts from a particular area such as measurement or measures of center.

development curriculum that discussed issues surrounding students' abilities to read graphs and the implications of data reduction as it relates to the increasing abstraction of what is actually being shown by graphs.

The Pedagogy Survey also was developed prior to the professional development curriculum. Its purpose was to reveal teachers' thinking about (a) the range of content encompassed by the term *statistics*, (b) ways that this content should be taught, and (c) particular uses of manipulatives and technology in statistics education. Teachers' perceptions of important statistics concepts shifted between the first two administrations of the survey. In the preinstitute administration, most of the responses about content appropriate for children could be categorized as "isolated content" or "simple process" items (e.g., graphing, probability, organizing data). Following the institute, most of the responses were categorized as "process" items (e.g., formulate questions, interpret data). Much of the emphasis during the summer institute was on modeling statistics activities through a four-step process (as noted earlier). Teachers seemed to have internalized the model, at least to the extent that their views changed about the statistics ideas appropriate for children to know.

Preinstitute instructional strategies dealt with data gathering, graphing, and computing statistical values. Following the institute, notions of appropriate strategies included data representation in a variety of modes, extended discussion with children, follow-up questioning by the teacher to help children develop meaning for interpretations of the data, and question formulation. These shifts reflect the instruction presented in the summer institute, so again, teachers seem to have internalized the goals of the institute (Bright, 1995).

The Pedagogy Survey did not provide insights into actual classroom behaviors. Through self-report, teachers indicated that believed that they made major changes in how they teach, believing that they did much better at questioning and encouraging inquiry. This observation also emerged in the open-ended response questions that were part of a later Teach-Stat Follow-Up Evaluation (discussed later). Again, classroom observations are needed to determine changes in teachers' implementation of instructional strategies. It is not clear that teachers have "moved" to the level indicated when they describe changes. However, it is clear that teachers believe that being involved with and implementing what they learned through Teach-Stat has had a strong positive impact on how they view teaching and on their students' responses to their teaching.

Professional Development Teams

Project staff noted that the experiences of the first-year teachers with their respective site faculty leaders appeared to be quite powerful for all involved. These experiences included participating in the first Teach-Stat institute, teaching with their students during the 1992–1993 school year, and participating as part of a professional development team (summer 1993) in conducting the second-year regional institutes. In order to try to capture summary impressions and reflections of these experiences, hour-long phone interviews were conducted in fall 1993 with 20 of the first-year teachers. A number of topics were addressed in these interviews, one being a discussion of their involvement in providing professional development for their peers during the second summer, particularly as at all sites "professional development teams" emerged as a model used. Stone (1994) characterized some of the teachers' views about these experiences in her summary of major themes from the interviews:

> Mrs. Robinson [a teacher] relays an insightful anecdote about the team's workings and teacher relationships with the university professor. She is, Robinson reveals, a kind of advisor that the teachers look to for leadership. But, this advisor also learned during the institute:
>
> > "She told us that she was so nervous about turning us loose to let us do this workshop because she wanted it to go well. But she said, after the first days, 'I am not nervous anymore—you guys are doing great.'" (p. 4)
>
> This experience was a first for Ms. Baker. And "it was a wonderful workshop." Part of the reason was the leadership of her team's professor who helped allay initial fears of teaching adults. The team met a couple of times in the spring for pre-planning and then met specifically to assign and plan lessons. What was interesting for Baker was that adults did have a different knowledge base than children, of course, but that one uses "the same strategies, just at a different level." (p. 7)
>
> This support was particularly evident during the second summer workshop when the team taught peers *Teach-Stat* for the first time. Some members like Mrs. Bates had conducted workshops in the past; other had not. This is how Bates puts this:
>
> > "We were very supportive of each other. We stayed right in the room . . . but not in an obvious way. Because we didn't want any one of us to look less professional than anyone else . . . We were very careful of each other's feelings yet at the same time, very, very professional."
>
> Part of this professionalism also was careful preparation of the workshop lessons: being prepared to do one's own lessons and knowing that others were doing as well. Finally Bates emphasizes two team elements. One is the idea that teachers never "talk down" when teaching—a model they appropriated from the Meredith College experience [the first-year institute in which they participated]. A second, significantly, is the relationship of the team's teachers to their "leader." She had become, it is explained, a kind of equal partner in *Teach-Stat* activities. (p. 10)

> Ms. Peterson says that the team members were personally selected by the professor who "did an excellent job getting a group of people who are the same and very different." Some are organizers, some are creative, some are perfectionists, some are not. Her job, as she sees it, is to add "creative spunk"; other present very well or demonstrate materials beautifully. And, they work so well together. (p. 22)

Project staff characterized this model simplistically as "take the course, apprentice teach the course with a 'master' experienced instructor, teach the course." In their work, Collins, Brown, and Newman (1989) noted that before schools appeared, apprenticeship was the most common means of learning and was used to transmit the knowledge required for expert practice (p. 453). Apprentices learn expert practice through participation in a process of observations/modeling, coaching, and fading.

> In this sequence of activities, the apprentice repeatedly observes the master executing (modeling) the target process. . . .The apprentice then attempts to execute the process with guidance and help from the master (i.e., coaching). A key aspect of coaching is the provision of scaffolding, which is the support, in the form of reminders and help, that the apprentice requires to approximate the execution of the entire composite skills. Once the learner has a grasp of the target skill, the master reduces (or fades) his participation, providing only limited hints, refinements, and feedback to the learner, who practices by successively approximating smooth execution of the whole skill.
>
> The interplay between observations [i.e., modeling], scaffolding, and increasingly independent practice aids apprentices both in developing self-monitoring and -correction skills and in integrating the skills and conceptual knowledge needed to advance toward expertise. (p. 456)

In looking at the earlier teacher comments we can see the apprenticeship model in operation. The observation/modeling component of this process is noted as a reference to the "Meredith College experience." Indeed, as the faculty participated together in the first-year professional development program (held at Meredith College), the notion of "modeling a process of teaching" was consciously addressed. When the teachers joined with the faculty leaders at each site in the second-year professional development program, they were given numerous opportunities to teach their peers. Several of the teacher comments point to the coaching role provided by the site faculty leaders. In addition, they highlight the important coaching role of their teammates during the workshops; further comments about the roles of teammates suggest that these teachers also functioned as models (sometimes successful and sometimes not so successful) for each other. Finally, faculty assumed the "fading" role, allowing the teachers to take on differing levels of responsibility and control during the workshops based on their readiness (which the teachers identified for themselves) to do so.

The project staff looked to this experience as a way to develop the Statistics Educators Institute related staff development manual. The goal was to try to capture in writing and activities much of what happened in the "on-the-job informal apprenticeship" experiences of the second summer institutes.

Statistics Educators—Developing Leaders

The benefit of a structure like MSEN is that it provides access to the state's school systems and assists in maintaining a consistent level of quality in the professional development programs it provides. However, North Carolina still lacks the capacity to provide high-quality opportunities for the majority of elementary teachers to deepen their subject matter knowledge and to continuously examine and modify their teaching practice. The Teach-Stat project sought to address the "capacity question" not only by providing professional development for a large number of teachers on a regional basis but, more importantly, by developing teachers (statistics educators) who can work with other elementary teachers in support of their learning statistics and about how to teach statistics and teach using statistics. This is one of the elements needed in building an infrastructure for professional development (Friel & Bright, 1997).

The final teams of statistics educators varied in composition: Some teams included only first-year teachers, some included a balance of first- and second-year teachers, and some included a few or no first-year teachers with the preponderance of second-year teachers. They were selected based both on their interest and on their potential ability to provide professional development to their peers. In cases where first- and second-year teachers were balanced, we found that teaming of a first-year teacher with a second-year teacher created a mentor/coach arrangement that seemed to support the second-year teachers in their initial experiences teaching other teachers. It was assumed that, in most cases, these teachers would work in teams of two statistics educators to provide such experiences for other teachers once they "graduated" as statistics educators.

Teachers selected for this opportunity participated in an additional week's professional development program that helped them explore staff development issues and ways to conduct a workshop. The Statistics Educators Institutes included content on adult learning, the change process, and statistics pedagogical content knowledge. Statistics educators completed 3 to 4 days of work prior to the Teach-Stat workshop they taught for third-year teachers; the remainder of the work was done as part of a "looking back" effort to reflect on what happened during the workshop.

As part of their participation, approximately half the statistics educators participated in a study (Frost, 1995) to investigate the effects of classroom teachers becoming Teach-Stat workshop leaders. They responded to three different instruments, and some also participated in interviews. These were completed at three points in time: at the beginning and again at the end of the Statistics Educators Institute, and after teaching the third-year Teach-Stat summer workshops.

This study is rich with information. For purposes here, the results suggest that staff development designs built on teachers becoming workshop leaders should provide special assistance to help teachers develop in this role over time. The following summary is relevant:

1. Opportunities to develop and/or demonstrate strong content knowledge in mathematics before becoming a workshop leader should be an important consideration in staff development.

2. Teachers' classroom experiences are valuable assets to their work as workshop leaders. Classroom experiences using teaching activities like those presented in workshops provide the workshop leader with "personal memory tapes" of the practical, as well as the pedagogical, issues related to the activities.

3. Teachers who become workshop leaders may need specialized assistance in conceptualizing effective staff development. The study suggests that workshop leaders progress through stages of growth in their conceptions about effective staff development; such stages can be used as "benchmarks" to assess readiness or potential of the teacher to serve as a workshop leader.

4. Teachers who become workshop leaders need opportunities to develop their own understanding of the nature of adult learners and of creating a climate conducive for adult learning. Further, there is a need to help workshop leaders explore pedagogical content knowledge related to teaching adults.

Teach-Stat Follow-Up Survey

The Teach-Stat project officially ended in August 1994. However, project staff have completed additional work in trying to assess the impact of the program. In spring 1995, a Teach-Stat follow-up evaluation was conducted. The purpose of the evaluation was to take a look at the overall effectiveness of the program's professional development activities. A modified version of a questionnaire (McMillen, 1995; Penta, 1995) developed by MSEN to assess impact of professional development projects was mailed to all Teach-Stat teachers (approximately 475 teachers including statistics educators). The questionnaire was in four parts; the first three parts were completed using a scannable answer sheet, and the fourth part was short answer response. The sections addressed (a) the amount and direction of change that teachers observed in students' behaviors over the period following the teachers' participation in the project, (b) the amount and direction of change in the teachers' personal teaching style over the same period, (c) the use and sharing of information and formal ways for sharing information from the project, and (d) the overall effects of the project.

The full report (Friel & Joyner, 1997) summarized data with respect to each of the sections noted. One area of special interest for the project staff was the ways teachers shared Teach-Stat with other teachers. The expectation was that statistics educators would be involved in more formal mechanisms such as presenting workshops or introductory sessions. One of the questions in the last section asked teachers to list workshops they had done in the last year (June 1994–May 1995). Statistics educators as a group conducted a number of 2-, 4-, or 6-hour Teach-Stat sessions, with a number of statistics educators also conducting longer workshops. Teach-Stat first-year and second-year teachers also have been involved in providing a number of introductory sessions for teachers at their school or in their school districts. Third-year teachers have done very few such Teach-Stat sessions, as they responded to this survey approximately 9 months (spring 1994) after they had participated in the program themselves and may have had few formal opportunities for such sharing. Respondents were also asked to describe one or more ways they worked individually with one or more small groups of teachers to support their learning to use Teach-Stat ideas. The responses from years 1–3 teachers offer a rich array of strategies. Table 4.1 provides several responses from both statistics educators and Teach-Stat teachers.

Several related comments have been highlighted in Table 4.1. These comments indicate that putting students graphs up outside teachers' rooms served to "invite" other teachers into the project. Stone (1994) made note of this as an important phenomenon in her report on teacher interviews. Site faculty leaders, in their reflections on the overall project, observed that the "graphs in the hall" phenomenon was a way of informing others about the project.

It is clear that Teach-Stat teachers, in general, have brought the project to other teachers. Much of this sharing has been informal and/or through short sessions. It is difficult to enumerate and then evaluate the impact of such sharing. The sharing does not equate with what the Teach-Stat teachers and statistics educators experienced, but there are now many resident resource teachers who can assist other elementary teachers in their efforts to teach statistics, one of the seven content strands of the North Carolina Standard Course of Study.

As was noted earlier, statistics educators have conducted a number of workshops for other teachers. Most often, workshops for school districts or schools have been of short duration. However, MSEN Centers each offered an additional 2-week workshop in the summer of 1995 that addressed learning statistics for elementary and middle-grades teachers. The site faculty

TABLE 4.1
Sample of Teach-Stat Follow-Up Survey Responses

Teach-Stat Statistics Educators	Teach-Stat Teachers
I have done a 10 contact hours workshop for my staff! I'm also a resource for them if questions arise.	I have given demonstrations of activities, helped set up purposeful graphing for curriculum fair, acted as an advisor to both individual and teams for graphing, statistics, and probability activities. My team set up a "Probability Fun Day" where students rotated to activity centers.
When my students do statistical investigations, we put their *graphs on the wall* outside my classroom. Curious teachers stop in and ask where I got the supplies (large graph paper & sticky dots). This has led to discussions about the T-S project, and several teachers have enrolled in summer workshops.	I worked closely with another teacher whose teaching situation is the same as mine. I shared materials & she did a unit in her class at the same time I did. That way we could share successes and frustrations.
During the fall semester at UNCC I taught a math course for pre-service teachers. I was able to share with the students there some activities to be used in the classroom with collecting, displaying, and interpreting data.	I worked with the Spanish teacher in setting up graphs of colors, etc.—using Spanish terms. I have worked with computer teacher in using Tabletop software to promote understanding of statistical concepts.
We have seven first grades. I have encouraged others to work with me in discovering many things about students and school. Topics: How many teeth lost in first grades? More boys/girls in 1st grade—Parking lot survey—What grade level wore the most glasses—left and right handed etc.	Last year I worked closely with another teacher on my grade level—once each week I would teach her class statistics for 40 minutes. This lasted the entire school year. It was a very memorable year with T-S.
I have presented workshops modeling & involving other teachers in investigations, shared my resources, & allowed teachers to observe my class in action.	Suggestions for using graphs, gathering materials & compiling/displaying info.
I have held a workshop for other teachers on ways to use T-S ideas in their classrooms. Also, I have presented ideas in a workshop at the State Math Conference.	Shared materials & activities with partner teacher & have seen increased use of manipulatives with her. Did activities with 3rd grade "buddy class" to increase awareness of collecting & displaying data.
Grade level meetings—I shared many ideas with my colleagues. One of the fifth grade teachers graphed the lunch choices for one month. Probability was an area that many teachers did not feel comfortable with. I shared ideas & teaching techniques related to this subject.	1. I made the materials I got available for check out so many children could benefit. 2. I had workshops for my grade level and 1st, 2nd, and 3rd grades. 3. I sent articles of statistical information or problems to colleagues.

(Continued)

TABLE 4.1
(Continued)

Teach-Stat Statistics Educators	*Teach-Stat Teachers*
I have worked with second and third grade teachers teaching the PCAI model. We have started with Posing the question with students, and then we have worked through the model. I team teach with these teachers, so after school each day we would get together to discuss problems and good experiences.	I have given ideas and experiments to fellow grade level teachers. I often had to explain terminology as well: mean, mode, range, median, stem and leaf . . .
I have helped conduct two workshops for teachers in our neighboring county.	*Putting graphs* our class has made in the hallway has sparked interest in this way of learning math.
I have tried to be a "resource person" on my second grade hall to answer questions about my *displayed class graphs* and to spark interest in other teachers to do T-S techniques. We graphed the number of books second graders read for the year. We polled other classes for investigations and compared various graphs between classes.	Weekly 1st grade team teacher meeting, *displays on school hall bulletin* boards, family math/science nights with parents and teacher coming from other schools to observe my classroom and math/science family nights, & student teachers.

leaders and/or statistics educators in each of the regions were responsible for conducting these workshops. Further, at three of the centers, special programs (from 10 hours to 2 weeks) with neighboring school districts were arranged that included workshops focused on teaching statistics; in each instance, statistics educators and site faculty leaders worked together to provide the professional development program.

We are unclear about how teachers develop sufficient depth in content knowledge and its applications to be able to serve as well-versed and skilled "consultants" in helping other teachers understand how to teach statistics. Informally, we have observed that teachers are successful at sharing their experiences and resources with other teachers, with many effectively serve in a coaching role as well. However, in our experiences with longer term (more than a day or two) workshops, we have found that statistics educators successfully "deliver" the Teach-Stat program, but often do not demonstrate the deeper insights into the subtleties of interactions with the content and pedagogy that also are needed in order to appropriately question or respond to workshop participants. Statistics educators comment that their participation in "student teaching" the summer workshop during the third year increased their understanding of the content in ways they had not expected. For those statistics educators for whom the third year was their second "student teaching" experience, we found and they found that they demonstrated greater insights and awareness with respect to the subtleties of content and pedagogy.

WHAT WE LEARNED

When the study of statistics is framed within the context of a process of statistical investigation and involves the use of relevant "hands-on" applications and activities, teachers and students quickly become engaged. Unlike much of traditional elementary school mathematics, teaching statistics within this framework provides for a much more open learning environment. No longer is there "one right way" to do mathematics with "one right answer" being the norm; questioning and exploration are encouraged and promoted. Professional development experiences that model such learning environments can be successful in helping teachers bring similar excitement and engagement in learning back to their students. We have found that the experiences gained with promoting inquiry learning while engaged in Teach-Stat are being transferred by teachers to other areas of teaching. The study of statistics and the process of statistical investigation is "extendable"; it integrates easily with other disciplines, providing for an interdisciplinary process that promotes consistency and connection across such subjects as social studies and science. Overall, individuals at all levels of involvement, primary-grade teachers to college teachers of statistics, learned from their Teach-Stat experiences and described these experiences as having impacted change in their respective classrooms.

ACKNOWLEDGMENTS

The development of this chapter was supported, in part, with funding from the National Center for Research in Mathematical Sciences Education (NCRMSE, funded by OERI), University of Wisconsin, and from the National Science Foundation, Project Teach-Stat. All opinions expressed are those of the authors and do not necessarily reflect the views of NCRMSE, NSF, or any other government agencies.

Teach-Stat Project Staff were: George W. Bright (UNC-Greensboro), Sarah B. Berenson (North Carolina State University), Theresa E. Early (Appalachian State University), Susan N. Friel (MSEN and UNC-Chapel Hill), Dargan Frierson, Jr. (UNC-Wilmington), Jane M. Gleason (North Carolina State University), M. Gail Jones (UNC-Chapel Hill), Jeane M. Joyner (NC State Department of Public Instruction), Robert N. Joyner (East Carolina University), Gary D. Kader (Appalachian State University), Nicholas J. Norgaard (Western Carolina University), Sherron M. Pfeiffer (Equals Consultant, Hendersonville, NC),

Mary Kim Prichard (UNC-Charlotte), Clifford W. Tremblay (Pembroke State University), and Elizabeth M. Vesilind (UNC-Chapel Hill).

REFERENCES

Berenson, S. B., Friel, S. N., & Bright, G. W. (1993a). The development of elementary teachers' statistical concepts in relation to graphical representations. In J. R. Becker & B. J. Pence (Eds.), *Proceedings of the Fifteenth Annual Meeting, North American Chapter of the International Group for the Psychology Mathematics Education* (Vol. 1, pp. 285–291). Pacific San Jose, CA: San Jose State University.

Berenson, S. B., Friel, S. N., & Bright, G. W. (1993b, February). *Elementary teachers' conceptions of graphical representations of statistical data.* Paper presented at the annual meeting of the Research Council for Diagnostic and Prescriptive Mathematics, Melbourne, FL.

Berenson, S. B., Friel, S. N., & Bright, G. W. (1993c, April). *Elementary teachers' fixation on graphical features to interpret statistical data.* Paper presented at the annual meeting of the American Education Research Association, Atlanta, GA.

Bright, G. (1995). *Final report of Teach-Stat pedagogy survey.* Greensboro, NC: UNC-Greensboro (unpublished).

Bright, G. W., Berenson, S. B., & Friel, S. N. (1993, February). *Teachers' knowledge of statistics pedagogy.* Paper presented at the annual meeting of the Research Council for Diagnostic and Prescriptive Mathematics, Melbourne, FL.

Bright, G. W., & Friel, S. N. (1993, April). *Elementary teachers' representations of relationships among statistics concepts.* Paper presented at the annual meeting of the American Education Research Association, Atlanta, GA.

Bright, G. W., Friel, S. N., & Berenson, S. B. (1993). Statistics knowledge of elementary teachers. In J. R. Becker & B. J. Pence (Eds.), *Proceedings of the Fifteenth Annual Meeting, North American Chapter of the International Group for the Psychology Mathematics Education* (Vol. 1, pp. 292–298). Pacific San Jose, CA: San Jose State University.

Brown, J. S., Collins, A., & Duguid, P. (1989). Situated cognition and the culture of learning. *Educational Researcher, 18*(1), 332–342.

Collins, A., Brown, J. S., & Newman, S. E. (1989). Cognitive apprenticeship: Teaching the crafts of reading, writing, and mathematics. In L. B. Resnick (Ed.), *Knowing, learning, and instruction: Essays in honor of Robert Glaser* (pp. 453–494). Hillsdale, NJ: Lawrence Erlbaum Associates.

Friel, S. N., & Bright, G. W. (1997). *Reflecting on our work: NSF teacher enhancement in mathematics K–6.* Lanham, MD: University Press of America.

Friel, S. N., McMillen, B., & Botsford, S. (1996). *Teach-Stat follow-up evaluation.* Chapel Hill, NC: UNC Mathematics and Science Education Network.

Friel, S., & Joyner, J. (Eds.). (1997). *Teach-Stat for teachers: Professional development manual.* Palo Alto, CA: Dale Seymour Publications.

Friel, S., Mokros, J., & Russell, S. J. (1991). *Statistics: Middles, means, and inbetweens.* Palo Alto, CA: Dale Seymour Publications.

Frierson, D., Friel, S. N., Berenson, S. B., Bright, G. W., & Tremblay, C. (1993, August). *Teach-Stat: A professional development program for elementary teachers (grades K–6) in North Carolina, U.S.A.* Paper presented at the meeting of the International Association for Statistical Education, Perugia, Italy.

Frost, D. L. (1995). *Elementary teachers' conceptions of mathematics staff development and their roles as workshop leaders.* Unpublished doctoral dissertation, University of North Carolina, Greensboro.

Gleason, J., Vesilind, E., Friel, S., & Joyner, J. (Eds.). (1996). *Teach-Stat for statistics educators: Staff development manual.* Palo Alto, CA: Dale Seymour Publications.

Gregorio, L. M., Vidakovic, D., & Berenson, S. B. (1995). *Teach-Stat evaluation report.* Raleigh, NC: Center for Research in Mathematics and Science Education.

Graham, A. (1987). *Statistical investigations in the secondary school.* Cambridge, England: Cambridge University Press.

Gravemeijer, K. (1994). Educational development and developmental research in mathematics education. *Journal for Research in Mathematics Education, 25*(5), 443–471.

Joyner, J., Pfieffer, S., Friel, S., & Vesilind, E. (Eds.). (1997a). *Teach-Stat for students: Activities for grades 1–3.* Palo Alto, CA: Dale Seymour Publications.

Joyner, J., Pfieffer, S., Friel, S., & Vesilind, E. (Eds.). (1997b). *Teach-Stat for students: Activities for grades 4–6.* Palo Alto, CA: Dale Seymour Publications.

Kader, G. & Perry, M. (1994). Learning statistics. *Mathematics Teaching in the Middle School, 1*(2), 130–136.

McMillen, B. J. (1995, March). *A follow-up evaluation: Teachers' ratings and reports of changes in classroom behavior after in-depth professional development programs.* Paper presented at the Annual Meeting of the North Carolina Association for Research in Education, Greensboro, NC.

McMillen, B. J., Botsford, S., & Friel, S. N. (1995). *Workshop evaluation.* Chapel Hill, NC: UNC Mathematics and Science Education Network.

Mokros, J., & Russell, S. J. (1995). Children's conception of average and representativeness. *Journal for Research in Mathematics Education, 26*(1), 20–39.

Moore, D. A. (1990). Uncertainty. In L. A. Steen (Ed.), *On the shoulders of giants* (pp. 95–137). Washington, DC: National Academy Press.

National Council of Teachers of Mathematics. (1989). *Curriculum and evaluation standards for school mathematics.* Reston, VA: Author.

National Council of Teachers of Mathematics. (1991). *Professional standards for teaching mathematics.* Reston, VA: Author.

Novak, J. D., & Gowin, D. B. (1984). *Learning how to learn.* New York: Cambridge University Press.

Penta, M. Q. (1995, March). *Overcoming practical difficulties in applying formal impact evaluation models to professional development programs.* Paper presented at the Annual Meeting of the North Carolina Association for Research in Education, Greensboro, NC.

Stone, L. (1994). *Teachers on Teach-Stat: A first report.* Chapel Hill, NC: UNC Mathematics and Science Education Network.

PART THREE

LEARNING STATISTICS

A Model-Based Perspective on the Development of Children's Understanding of Chance and Uncertainty

Jeffrey K. Horvath
Richard Lehrer
University of Wisconsin–Madison

Statistics is the science of modeling the world through the theory-driven interpretation of data (Burrill & Romberg, chapter 2, this volume; Schwartz, Goldman, Vye, Barron, and the Cognition and Technology Group at Vanderbilt, chapter 9, this volume; Shaughnessy, 1992). Models of chance and uncertainty provide powerful cognitive tools that can help in understanding uncertain phenomena. In this chapter, we consider the development of children's models of chance and uncertainty by considering their performance along five distinct, albeit related, components of a classical model of statistics: (a) the distinction between certainty and uncertainty, (b) the nature of experimental trials, (c) the relationship between individual outcomes (events) and patterns of outcomes (distributions), (d) the structure of events (e.g., how the sample space relates to outcomes), and (e) the treatment of residuals (i.e., deviations between predictions and results). After discussing these five dimensions, we summarize and interpret the model-based performance of three groups as they solved problems involving classical randomization devices such as spinners and dice. The three groups included second graders (age 7–8), fourth/fifth graders (age 9–11), and adults. We compare groups by considering their interpretations of each of these five components of a classical model of chance. We conclude by discussing some of the benefits of adopting a modeling stance for integrating the teaching and learning of statistics.

MODELING CHANCE AND UNCERTAINTY

How might an "expert" model of chance and uncertainty account for the outcomes observed in tasks involving spinners or dice? What components would such a model contain? In this section we elaborate on the cognitive status of five components of the classical model of chance.

Distinguishing Between Certainty and Uncertainty

Statistics fundamentally deals with measuring degrees of uncertainty with probabilities, likelihoods, confidence intervals, and the like. Even 4- and 5-year-old children understand the concept of uncertainty in some systems. For example, Kuzmak and Gelman (1986) found that 4-year-old children could distinguish between a situation of certain outcome (i.e., determining the color of a marble dropped out of the bottom of a thin transparent tube with marbles stacked one on top of another) and one of uncertain outcome (i.e., determining the color of a marble dropped out of a metal cage where marbles were mixed by rotating the cage until one marble randomly dropped out of a small opening). However, it is sometimes problematic, even for experts, to decide if a system should be treated as fundamentally predictable or unpredictable. For example, in a chaotic dynamic system (e.g., weather), a simple set of rules completely determine the course of events. However, uncertainty creeps in during measurement. Small, unmeasurable fluctuations in starting states can produce unpredictable changes in future events. Situations like these are more appropriately viewed as uncertain, even if the system is described by a simple set of determinant rules. Hence, uncertainty is not a fundamental property of a phenomenon but, rather, is a conception that is situated in a context (see also Metz, chapter 6, this volume.)

The Nature of an Experimental Trial

Models of probability assume a means of marking or classifying each event in order to combine and compare sets of events. According to Stigler (1986), "It [is] an essential precondition for probability-based statistical inference that in order to be combined, measurements should be made under conditions that [can] be viewed as identical, or as differing only in ways that [can] be allowed for in the analysis." A trial is an instantiation of an experiment (e.g., the spin of a spinner) that yields a public outcome (e.g., the resultant number). Two instantiations of an experiment can be compared if those two instantiations are identical with respect to some key task structures (Fischbein, Nello, & Marino, 1991). For example, two spins of a spinner, one spun clockwise and one spun counterclockwise, would be considered two trials of the same experiment in most contexts because both cases are identical with respect to the key task structure—uncertain outcome. Alternatively, two spins of a spinner, one spun gently and the other spun vigorously, might not be considered two trials of the same experiment if the gentler spin is viewed as generating a less random result. Different understandings of the nature of experimental trials will result in different stances about what events should be considered equivalent or exchangeable.

Relationships Between Simple Events and Distributions

Although individual events may be highly unpredictable (e.g., the result of a single spin of a spinner), global patterns of outcomes are often predictable (e.g., for a spinner with three congruent regions, one would expect each outcome would occur with approximately equal frequency). The tendency for global structure emerging from locally unpredictable events is often referred to as the "law of large numbers" (see Stigler, 1986, pp. 64–70). This "law" suggests that as the number of events gets very large, the pattern of results will converge to some distribution, a distribution dictated by the structure of the events.

Again, even for experts, it is sometimes difficult to know if it is reasonable to expect the law of large numbers to apply to a system. Recognizing that the law of large numbers applies to the results from a classical randomization device such as a die or a spinner will add significant predictive and explanatory power to a model. However, the same reasoning may not apply to chaotic systems.

Structure of Events

Classical models of chance rely on the concept of a sample space—a model of the aggregate structure of outcomes inherent in the randomization device. Understanding of a sample space requires orchestration of several cognitive skills. For example, consider the task of predicting the outcomes of spinning two spinners with three congruent regions labeled 1, 2, and 3 where the outcomes will be the sums of the two results (i.e., the numbers 2–6). First, one must recognize that there are different possible ways to achieve some of the outcomes (e.g., there are two ways to get a sum of 3, a 1 on the first spinner and a 2 on the second spinner and vice versa). Second, one must have a means of systematically and exhaustively generating those possibilities. Third, one must map the sample space onto the distribution of outcomes. We discuss later some of the reasons why some of these principles may not be transparent to children.

Treatment of Residuals

Models are not mere copies of phenomena. They are abstractions of key structures and relationships present in the phenomena being modeled. Because of this, there will be residuals (i.e., difference between predicted and actual results) between the model and the phenomena being modeled. In the classical model, batches of data can be viewed as approximations of the (ideal) sample space. The sample space provides one with a principled way of thinking about and accounting for sample-to-sample variation and residuals. Without constructs like the sample space, chance deviations may be interpreted as evidence of an ill-fitting model (Lehrer, Horvath, & Schauble, 1994). Even given constructs like a sample space, however, judging the adequacy of fit of a statistical model is often not trivial because it requires judgment about the magnitude of deviation that should be taken as evidence of an ill-fitting model. How a person understands residuals will have a large effect on the person's model-based reasoning about an uncertain phenomenon.

Throughout this chapter, we treat these five dimensions—distinguishing between certainty and uncertainty, the nature of an experimental trial, relationships between simple events and distributions, structure of events, and treatment of residuals—as if they were separate strands of understanding chance and uncertainty. In reality, however, the five strands are often interrelated. In fact, in many ways they form a natural progression in understanding the dimensions of chance. For example, understanding the relationship between simple events and distributions presupposes an understanding of the nature of experimental trials. Understanding the role of the sample space in chance investigations presupposes an understanding of some relationships between simple events and distributions.

In summary, even simple models of chance and uncertainty involve the orchestration of several interrelated components of reasoning, each of which can pose significant intellectual challenges even to statisticians. We next contrast the performances of children and adults as a means of investigation transition within and relationships among these five components of reasoning.

STUDIES OF PERFORMANCE

We turn now to studies of children's and skilled adults' ideas about chance and probability in the context of simple randomization devices such as dice and spinners. We chose to study participants' explanations of the relationship between observed and expected outcomes because these devices provide contexts in which the operation of chance is fairly obvious. It is also generally agreed that statistics based on classical probability theory provides a good model of the phenomena produced by these devices. Consequently, we could focus on children's appreciation of models of chance, rather than on other factors, such as a potential competition between chance and cause, etc. (see

Derry, Levin, Osana, & Jones, chapter 7, and Schwartz et al., chapter 9, this volume).

The first study was conducted in two second-grade classrooms of approximately 25 students each (ages 7–8). The classrooms were located in two different suburban Midwest public elementary schools in the same school district. Classroom teachers were participating in a project of mathematics reform emphasizing the development of mathematical reasoning by building on students' informal knowledge of space and number (Lehrer, Jacobson, Thoyre, Gance, Horvath, Kemeny, & Koehler, in press). The second-grade students participated in experiments involving one or two 6-, 8-, or 12-sided dice. Students made predictions, generated results, and justified the relationships between expected and obtained results. They made extensive use of bar graphs for recording results and sample spaces. We relied on field notes, audiotape, and videotape to conduct our analysis. Our discussion focuses largely on the performance of the children in one of the classrooms (classroom A). We report findings from the second classroom (classroom B) only when they differed from those of the first classroom or when they provided an elaboration of evidence of second graders' reasoning about these tasks.

In a second study, a pair of fourth/fifth-grade students and a pair of statistically sophisticated adults both participated in an activity titled the Spinner Sums Investigation (SSI) designed by the California Department of Education (California Assessment Program, 1991). The activities in the SSI involved one or two spinners with three congruent regions labeled with the numbers 1, 2, and 3. Subjects were asked to predict the results of a given number of trials, generate results, and explain any residuals (differences between predictions and results). We relied on field notes and videotape to conduct our analysis. Because we have contrasted the performance of these two pairs (fourth/fifth graders and adults) elsewhere (Lehrer, Horvath, & Schauble, 1994), we present results of this second study only in enough detail to serve as comparative benchmarks for the results of the first study.

The sample size for the fourth/fifth graders (2) and adults (2) was small in relation to that of the second graders (25). However, the intent of the study was to begin to develop a model of transition in statistical understanding of chance and uncertainty through case studies. Such a model, once developed, can be tested further with larger samples.

MODEL-BASED REASONING ABOUT CHANCE IN THE SECOND GRADE

Overview of Tasks and Problems Posed to Students

One classroom of second-grade students (classroom A) participated in experiments involving dice over a period of 7 days. Each day's activities lasted approximately 45–60 min. Throughout the 7 days, the children participated in three different types of activities. In the first activity, pairs of children predicted the most and least frequent outcome when rolling one or more dice, generated results for 30 or more trials, and represented results with a bar graph. They did this for one and two six-sided dice and one and two eight-sided dice. During the second activity, children generated sample spaces (i.e., the number of ways of obtaining each possible outcome) for the various die combinations just noted and then represented the frequency of each possible outcome with a bar graph. For each of these first two types of activities (predicting outcomes and generating sample spaces), the students worked in pairs. After each pair of students had generated their results or finished describing their sample spaces, the teacher gathered the class, displayed each pair of students' graph on the front board, and led a group discussion of the results. For the third activity, the entire class generated one large composite bar graph of the results each student obtained when rolling two six-sided dice. The teacher polled each student for his or her individual results and represented the results on one large bar graph displayed on an overhead projector.

The other classroom of second-grade students (classroom B) participated in activities very similar to that of classroom A (e.g., making predictions, generating, notating, and discussing results). They also participated in two additional activities that were identical in structure to the other tasks previously described (e.g., predict, generate, and explain results of experiments) but used different dice: one involving a single 12-sided die and one involving a single nonstandard 6-sided die (four of the sides were labeled 2 and the other two of the sides were labeled 6).

In this section, we use the five components of the classical model described previously as an interpretative framework to describe children's understanding of chance. The range of results and observations we report are representative of the children in these two classrooms.

Distinguishing Between Certainty and Uncertainty

All of the second graders clearly distinguished rolling dice from other more determinate actions in their classroom like writing, walking from one place to another, or doing arithmetic problems. Nevertheless, many children did not think of rolling dice as completely random; some children predicted events based on a past history of "lucky" numbers or believed that their wishes about an expected value would somehow influence, albeit not completely determine, the trajectories of the dice. However, beliefs about lucky numbers and partial telekinesis usually did not endure long after the children's first opportunities to collect data about rolling dice. These observations are consistent with other research supporting the finding that children of this age (7–8 years) can and do distinguish certainty and uncertainty for simple devices such as the dice employed in this study (e.g., Kuzmak & Gelman, 1986). Nevertheless, a substantial minority of the children did not initially view dice as completely subject to chance.

The Nature of an Experimental Trial

On the first day of experimentation in classroom A, the students noticed that they were not all tossing two six-sided dice the same way. Some were tossing both dice at the same time, some were "palming" the dice, and some were tossing one die and then selectively tossing the second die at that first die, causing a collision. This led to a class discussion about the uncertainty of the outcomes associated with the various methods. The conversation focused on the way that one particular student, S1, was rolling his dice. S1 had predicted that he would get more 12s than any other result because 12 was his lucky number. When he noticed that he was not getting very many 12s, he began throwing the dice sequentially. He threw the first die. If it resulted in a 6, he would throw the second die normally, hoping for another 6. If the first die did not result in a 6, however, he would throw the second die at the first, hoping to change it to a 6, thereby giving himself a chance at a 12. He argued that regardless of the way that he rolled the dice the outcomes were still uncertain and, hence, his methods were acceptable.

After listening to this argument, about half the class thought that tossing the dice different ways had no effect on the outcomes, accepting S1's argument that any method was acceptable for which the final outcome was not completely determined. The other half of the class, however, disagreed. They agreed with S1 and other classmates that uncertainty of outcome was necessary. However, they also suggested that in order for the successive throws (trials) to be compared, each toss of the dice should be "fair." As the following transcription shows, these students thought that S1's technique of selectively trying to alter the result of the first die changed the nature of the experiment:

S2:   That is cheating. [A reference to S1's method of biasing the outcomes by selectively hitting one die with another after checking on the outcome of the first die.]

S3:   It is like you are almost picking it up and doing it.

S2:   It is like inventing a new dice game or something.

These students believed that bias changed the nature of chance in such a way as to make it a different task (Fischbein, Nello, & Marino, 1991). To them it was "cheating" or "like inventing a new dice game." The experimental events were no longer comparable, and this seemed unacceptable to them.

Each of the students seemed to have a strong sense that something about the trials needed to remain constant in order for them to be comparable (i.e., "fair"). There was not, however, a consensus about exactly which features of the task needed to remain constant. For some students, all that was necessary was some uncertainty in outcome. Other students, however, also demanded that there be no biasing of the outcomes.

Eventually, the class agreed that for all future experiments they would roll their dice out of a cup. Some of the students recalled doing similar things when playing games with dice. It was agreed that rolling dice out of a cup ensured uncertainty of outcome, and that if both dice dropped out of the cup at the same time, there could be no biasing. By adopting this convention, the class defined the nature of trials for the remainder of the tasks in which they engaged. They then felt free to compare, contrast, and combine their future results.

Relationships Between Simple Events and Distributions

All of the children understood that simple events (i.e., single outcomes) were highly uncertain and unpredictable for the classical randomization devices that they were working with. Their reasoning about the predictability of patterns of events was much less uniform, however. Although a few children maintained throughout the investigation that patterns of outcomes were just as uncertain and unpredictable as any single outcome, most of them believed that patterns were more predictable than single outcomes. Their understanding of this relationship developed as they had more experience with the tasks and varied in different contexts.

Basing Predictions on Past Experience. One of the first contexts during which the relationship between local outcomes and patterns of outcomes was explicitly raised in classroom A (it was a tacit assumption of the teacher's request that children predict outcomes after a number of trials) occurred on the second day of the investigation. The children were asked if they could predict what their graphs would look like after 50 trials, given their own graphs of approximately 20 trials with one six-sided die. A number of children (although not all) believed that the pattern of results was predictable and based their predictions on past results such as the first 20 outcomes:

Investigator: Okay . . . what is it [the shape of the distribution after 50 trials] going to look like? . . . If you rolled it [the die] 50 more times, what would the shape of that [the distribution] look like? . . . Who can come and draw for me . . . S4? What do you think the line would look like? . . . Okay, so S4 says that the graph will look like this [a translated version of her original graph of results].

S4 believed that she could predict what the distribution would look like in the long run. She based her prediction for that distribution on the results that she had just generated. She used the shape of the distribution describing her recent results to describe the distribution of her predicted distribution of future results, simply translating that distribution to account for the larger number of trials.

Over the course of the first few days of investigation, most students developed a belief that global patterns of events were more predictable than local outcomes and that those global patterns were informed by past experiences, just as S1 did. During these first few days, when the students primarily worked independently or in pairs making predictions and generating results, very few of them independently made any connections between any relationship between patterns and individual outcomes with sample spaces for those domains. In the latter days of experimentation, a few contexts were introduced that increased the frequency with which the children related the predictability of patterns with the structure of the domain (i.e., the sample space).

Contextually Supported Reasoning About the Role of the Sample Space.   On the third day of experimentation in classroom B, the investigator posed two questions to the class concerning outcomes when rolling the single nonstandard six-sided die (four 2s and two 6s). While holding the nonstandard die in front of the class, he asked, "How certain are you of the result when I roll this single die? How certain are you of the shape of the graph if I were to roll this single die 1,000 times?" The students debated the specifics of the shape of the eventual graph (e.g., would the bar for 2s be twice as high as the bar for 6s or would it be three times as high?) but were all confident that the 2s bar would be significantly higher than the 6s bar. They were "really really" sure that they could predict the shape of the distribution of results but not at all sure that they could predict the result of any one toss. Again, the children were much more confident about how the distribution of results would look than about the outcome of any single trial.

What this context did was make the structure of the sample space much more evident to the students while they were reasoning about the predictability of the distribution of outcomes. The single die, with its four 2s and two 6s, embodied the structure of the sample space. Discussing the predictability of the results—individually and collectively—with the sample space made more evident resulted in the entire class feeling very confident about the predictability of the pattern of results while still being uncertain about the result of any single outcome.

On the sixth day of the investigation in classroom A, the children were asked to participate in a similar but much less spontaneous and much more supportive activity in which the issue of the relationship between pattern of results and individual results was also addressed. They were asked to participate in a collaborative experiment involving two six-sided dice. The structure of the task was identical to the earlier one involving two six-sided dice (i.e., generating sums of the two dice). This time, however, the students did not record their own individual results on a bar graph. Instead, the class generated a composite graph of the combined results of all the students' trials. The teacher went around the room and asked each student to generate a single result, which she then recorded on a bar graph displayed on an overhead projector at the front of the class. After finishing the first epoch, the teacher then polled each student for another (single) result, recording those results on the same bar graph. The results of each die toss for a single epoch were recorded in a different color so that, for example, the first epoch's results were depicted in red, the second epoch's in blue, and so on.

During this activity, the students got to see many more entries displayed on a single graph than they had seen during any of their previous experiments and were supported by the conversation of the teacher and fellow classmates. The format of the activity afforded a discussion of how the individual results and small sets of results (i.e., small aggregates of individual outcomes such as the results of each epoch) related to the distribution (or potential distributions) of outcomes. During the course of the experiment, the teacher periodically paused to ask the class what they thought of the task so far and what they would predict for the future:

Teacher: Okay. That is 33 rolls so far and when you guys go rolling, just you and your partner, often you weren't rolling too many more than about that much. . . . So what I am going to do now, then, is I am going to [graph the next set of results] in another color for you . . . [and see what happens] as we roll more times. . . . What do you notice happening?

S5:   [The ones that were lower] on the red . . . are getting higher.

T: What do you mean by that, the ones that were a lot lower?

S6:   [Refers to entries with low frequencies on the first portion of the graph]

I: S6 says it is beginning to look more like steps.

T: When we go through our next ones, we will do these in green. . . . What do you think might expect to happen?

[S7 describes a step-like graph with the center numbers having higher frequencies than extreme numbers.]

In this context, the students began to expect the distribution to have a particular shape; they expected it to look more and more like "steps" (i.e., central values occurring more often than extreme ones) as the graph grew. They actively cheered for any result that moved the graph closer to a step graph. Given this enthusiasm, it seems reasonable to say that they thought that the distribution was predictable, that it would look like a step graph.

The Role of the Notational Assistance. What effect did the introduction and use of the bar graphs have on the children's reasoning? Most obviously, they acted as external placeholders for information, reducing the cognitive burden for the children. But what information did the children get out of the graphs? The tasks the children were engaged in, predicting the most or least frequent results, did not lend themselves to making the distinction between a sample space based on combinations and one based on permutations obvious. The distribution of the space of combinations and that of permutations are both roughly triangular-shaped distributions with central values occurring more frequently than extreme values. The difference between viewing the sample space as being comprised of combinations or permutations only becomes relevant when making some very fine discriminations. For example, it would be very relevant when trying to explain why a certain experiment with two six-sided dice resulted in more 3s than 2s.

The bar graphs served best to highlight the general shape of the distribution of frequencies for each outcome. The children referred to the sample space graphs (based both on combinations and permutations) for the two-die cases as looking like "steps" or "mountains" (i.e., a step function with central values being higher than extreme values) and to the sample space graphs for the single-die cases as looking like "wagons" or "waves" (i.e., flat, with each value being roughly equal), emphasizing the fact that they were attending to the overall shapes of the distributions.

The specific values for the number of "ways of getting each result" depicted in the sample space graphs (e.g., the number of ways of generating an outcome of a 3) were not as salient to the children as was the overall shape. The tasks and the conversations surrounding them were designed to engage the children in thinking about relationships between sample spaces and outcome spaces. Given the nature of the task, however, small samples of results (like those generated by the children) were highly variable. As such, subtle differences between, for example, the number of times a 2 was tossed and the number of times a 3 was tossed were not very significant, and the notational system did little to resolve the issue of whether the two cases should indeed be treated differently.

The bar graph notation highlighted the general shape of the distribution and magnitude of results. It made the absolute differences between particular results much less salient. In terms of local/global, the bar graph notation encourages children to notice relationships at a global level (i.e., distributions) rather than at a local level (comparing individual outcomes).

The level of understanding of the relationship between individual results and global patterns varied widely over time and across contexts. Some students believed that the patterns of results were as unpredictable as individual results. Most, however, thought that patterns were much more predictable and initially based their predictions of patterns on past results. A pair of activities, however, made the relationship between the structure of the tasks (i.e., the sample spaces) and the predictability of the patterns of results clearer to the children. In classroom B, when the sample space was made evident to the students (they could simply look at the nonstandard die and count the relative number of occurrence of each possible outcome), the class was "really really" sure of the predictability of the global but not very sure at all of the predictability of any local outcome. When classroom A generated a composite graph with numerous entries, and the activity was supported by the conversation led by the classroom teacher, the children demonstrated an expectation that the global was predictable (cheering results that brought the graph closer to a "step" graph) and stated that they expected future results to bring the graph ever closer to an ideal step graph. The different task contexts seemed to greatly influence the degree to which the students saw any relationships between local outcomes and patterns of events.

In order to fully understand how and why patterns of results are more predictable than individual results (for classical randomization tasks like these) the children needed to understand the concept of a sample space and how it informed such relationships. The children never understood the role of the sample space without significant support and, hence, never completely understood the reasons why patterns were more predictable than simple outcomes. This is discussed in the next section.

Structure of Events

In order to begin to understand how, or even if, the second-grade children thought about relationships between the structures of the tasks they were engaged in and the outcomes they were generating, the students were asked to participate in a series of activities in which these relationships would be made apparent. The activities involved the generation of sample spaces (e.g., listing all the ways each possible outcome could occur) and the consideration of how those sample spaces related to outcomes. In order to further assist the children in their thinking about these issues, two forms of assistance were provided, one notational and one conversational. The notational assistance was a bar graph used to represent the number of different ways of generating each outcome. The conversational assistance took the form of classwide conversations led by the teacher wherein the conversation of the other students and that of the teacher served to support the reasoning of the individual students.

Combinations and Permutations. On the fourth day of experimentation in classroom A, the class was asked to generate all the "ways you could get" (the distinction between combinations and permutations was not initially made) each of the possible outcomes when rolling two eight-sided dice. The activity was meant to try to make the relationships between the sample space and the outcomes clearer to the students. The students were asked to make a chart listing each of the "ways" they could get each result and then create a bar graph (like those used to depict results) representing the number of ways of generating each result.

After all students had generated their charts and graphs, the class reconvened and discussed the various results, taking advantage of the conversational support provided by the teacher and class. A lengthy discussion ensued about whether or not order mattered when listing a "way to get" a result. They debated, in other words, whether they should count all permutations or just combinations. The structure of the sample space for this task is dictated by the relative distribution of the number of "ways" to get each result. For example, if one only considers combinations, then there are three times as many ways of achieving a 6 as a 2 when rolling two six-sided dice (3 vs. 1). If one considers permutations, however, there are five times as many ways of achieving a 6 as a 2 (5 vs. 1). In the following discussion, the investigator is asking the class how many ways there are to roll a sum of 6 using two six-sided dice:

Investigator: How many ways all together are there to make a 6?

S1:   Three.

S2:   Five.

I: Some of you say 3 . . . and some of you say 5 . . . Do you think it is 3 or do you think it is 5?

[the class is split, most thinking 3 but some thinking 5]

I: Okay. Someone who thinks it is 3, explain why. S3?

S3:   Um, because . . . this is [just] two of the ways [of] writing the numbers. It would be the same if you wrote this one this way [if you wrote 1 and then 5 instead of writing 5 and then 1]. It would be the same way.

Some of the children (e.g., S3) reasoned from the commutative property of addition (e.g., $1 + 5 = 5 + 1$) to suggest the equivalence of various combinations (e.g., 1,5 and 5,1). Others, such as S4 in the following, saw the task as different from other commutative ones, such as addition, and viewed differently ordered outcomes as separate.

[Instructor asks S4 why she thinks that there are 5 ways to make a 3 with two six-sided dice]

S4:   Because, okay. These two ways [1/5 and 5/1], maybe they could be the same numbers if you put the difference . . .

I: You mean the five and one, and one and five, could be the same numbers but they are different [i.e., they count as a different "way to get" the number]?

S4:   Yes.

I: How are they different?

S4:   [They switch the] numbers.

I: They switch which side is producing a one and which one is producing a five? Okay.

The debate over whether or not order mattered was impressive to see in second graders. Thoughtful consideration of whether to consider permutations or just combinations indicates a concern for the structure of the sample space, although children never reached consensus about this issue (the difference does not affect the "step-like" nature of the resulting distribution, only the relative heights of the steps). For the purpose of this discussion, we refer to both combinations and permutations as "possibilities" or "ways of getting" (as that was the vocabulary adopted by classroom A). The second graders unsystematically generated the possibilities when rolling two eight-sided dice. Figures 5.1 and 5.2 show a sample chart and bar graph created by one of the students.

Comparing Result Graphs To Sample Space Graphs.   The day after generating charts and bar graphs of the sample space, the teacher in classroom A led a discussion comparing the graphs depicting the sample spaces with those depicting the results of the previous experiment with two eight-sided dice. The goal of the discussion was to see if a direct comparison of the information in the two types of graphs (e.g., shape, most frequent result, etc.) would prompt the students to consider the relationships between sample spaces and outcomes. The teacher placed particular emphasis on how the "number of ways" of achieving each outcome (as depicted on the sample space graphs) might relate to the frequency of actual results if the students were to generate a set of outcomes. Initially, all the students noticed about the two types of graphs was that they looked similar, that is, they both were step-like. However, with much conversational support from the teacher and the reification of the sample spaces and outcomes using the bar graph notations, the children were occasionally able to articulate some of the relationships between the two types of graphs (i.e., outcomes and sample space):

Teacher: Okay . . . S6, you were saying that they [i.e., extreme values] don't look quite the same [as central values], they don't go as high? . . . Why do you think that is? Why do you think this graph [the extreme values] doesn't go as high as our other ones [height refers to frequency of outcome]?

S6:   [it depends on] how many ways you can do it, so maybe one way would be . . . 9. So you could get 9 more than 6 because you have a lot of more ways that [you can get] that number.

numbers you could get rolling two 8-sided dice	ways to make the number using two 8-sided dice
2	$1+1=2$
3 $1+2=3$	$2 + 1 =3$ $1+ =3$
4 $3+1=4$	$2+2=4$ $1+3=4$
5 $2+3=5$ $4+1=5$	$3+2=5$ $1+4=5$
6 $1+5$	$3+3=6$ $5+1=6$
7 $1+6=7$ $3+4=7$	$6+1=7$ $4+3=7$
8 $3+5=8$	$4+4=8$ $7+1=8$
9 $4+5=9$	$8+1=9$ $5+4=9$
10 $8+2=10$ $4+6=10$	$5+5=10$ $7+3=10$ $6+4$
11 $6+5=11$ $5+6=11$	$8+3=11$ $5+6=10$
12	$6+6=12$ $5+7=12$
13 $6+7=13$ $5+8=13$	$8+5=13$ $7+6=13$ $8+4$
14 $6+8=14$	$8+6=14$ $7+7=14$
15	$8+7=15$ $7+8=15$ $6+8=13$
16	$8+8=16$ -

FIG. 5.1. Chart of all the "possible ways" of getting each sum using two eight-sided dice.

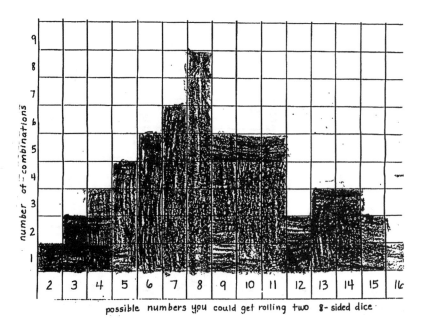

FIG. 5.2. Bar graph depicting the number of ways of getting each sum using two eight-sided dice.

The students, with much support, understood that the sample space graphs contained information about how many different possibilities there were for getting each outcome and the empirical graphs (i.e., outcomes) contained information about actual outcomes. They also understood that the distribution of results for their outcomes was related to the distribution of results in the sample space. In the context of the group discussion with the notations in front of her, S6 explained the phenomenon in question (central values occurring more frequently than extreme values) in terms of the structure of the sample space.

The students' understanding of these relationships was very fragile, however. When the conversation stopped and the graphs were no longer right in front of them, the students no longer understood how the sample space related to the outcomes. After discussing the relationships, the students were asked to make some predictions about what their graphs would look like after rolling a single eight-sided die 24 times. Very few of them incorporated any information about the number of possible ways of getting each outcome. Instead, they reverted to reasoning about empirically obtained results:

Investigator: Okay. I just asked S7 to explain his graph. S7 has a graph that has 2 once and 3 once and 4 twice and 8 about, how many times is 8?

S7: Eight.

I: About 8 times? So why did you pick so many 8s? [S7 doesn't know. He shrugs.]

I: Okay. I am looking at S8's graph. And S8 has 6 appearing most often and then 5. S8, why did you think it would come out that way?

S8: Well, I usually roll 6s.

I: Oh, you usually roll 6s? Okay. Thank you.

Without the support of the conversation (from the teacher and fellow classmates) and notation (i.e., the set of graphs displayed at the front of the room), many children seemed to lose track of the relationships that they had just been discussing and fell back on haphazard prediction strategies such as S7's or on previous experience as did S8, not on any relationships between the sample space and the outcome space.

Collaborative Graphing. Returning to the activity of day six in classroom A, generating a composite bar graph of the entire class's results, we see more evidence that when the second graders were attending to the notational tools and were being supported by the classroom conversation, they were better able to reason about the relationships between the sample space and the outcome space. We start this discussion with the teacher probing the class to see what they think about the shape of the graph that they are constructing. The class begins to construct the composite graph as described in the section on their understanding of the relationship between individual results and collections of results, and the teacher probes the children for their reasoning about the role of the sample space in explaining the results that they are generating:

Teacher: What do you think about the other numbers? What has been happening to the other numbers so far?

S9: More 6s, more 8s to go with the more 7s.

S10: Because usually the holes start filling up.

S11: There was only one way to get a 2 with this pair of dice and there is more ways to make 6 and 7.

T: What happens the more times we roll something? Over a long period of time, what starts to happen? . . .

S12: The middle starts to become the highest.

S13: They have more ways of doing it in the middle . . . more ways.

T: With that number?

S14: Yah.

Investigator: S14 said there are more ways to get the numbers 6, 7, 8.

S14: [There are more ways to get the numbers in the] middle **than** on the outside.

In this group activity, with the support of the notation being generated by the teacher and the conversation of the class as a whole, the students were able to incorporate information about the sample space into their discussion of the outcomes. They correctly reasoned that the central values were occurring more often than extreme values because "There was only one way to get a 2 with this pair of dice" and "There is more ways to make 6 and 7." In this context, the students were able to reason about the relationships between sample space and outcomes. The difference between the students' independent performance (when they did not make use of any relationships between sample space and outcome) and their assisted performance (when they did) demonstrates that an understanding of these relationships is with the students' zone of proximal development (Vygotsky, 1978).

In summary, when experimentation began in both second-grade classrooms and the contexts for the activities centered around making predictions about specific results or sets of results, most students did not reason about the structure of the tasks they were engaged in or relate the sample space to outcomes in any way. When the sample space was deliberately introduced into the students' investigation of chance and uncertainty and two forms of assistance were introduced, a notational system for reifying sample spaces and the support of the conversation of the teacher and the class as a whole, most students were able to at least recognize and understand the concept of a sample space and infer some of the relationships between a sample space and outcomes. The context prompted the students to think much more deeply about the structure of the events they were generating than when they had been working independently on their own smaller graphs. They debated at length about whether it was important to consider combinations of die results (where order did not matter) or permutations of results (where order did matter). When the context changed and the forms of assistance were removed, however, most students reverted to reasoning about previously obtained empirical results, ignoring information about the structure of the events.

There is evidence, then, that the second-grade students began to understand the mathematical/theoretical relationship between the sample space and the probabilities of outcomes. Most of the students could accurately identify the information contained in empirical and sample space graphs, and in some situations they would relate the information between the two when appropriate forms of assistance were provided. Without these forms of assistance they were less able to reason about relationships between individual results and patterns of results.

Treatment of Residuals

The activities that the students engaged in were designed to allow the students to create and use models of chance as they participated in a series of investigations. As mentioned previously, residuals are indicators of the goodness of fit of a model to a phenomenon. Consequently, how the students understood and treated residuals would have a significant impact upon the generation and evolution of their models.

Reasoning About Individual Results. When experimentation began in both classrooms, the students' predictions were not in any way based on a model of the phenomena they were investigating. A conventional model-based prediction for an experiment involving classical randomization devices such as spinners relies on the concept of a sample space to inform predictions. Those predictions are typically based on the distribution of results in that sample space. When experimentation began, these second-grade students' predictions and explanations of outcomes dealt primarily with single outcomes, such as which number would "win." Their reasoning was informed by ideas like "lucky numbers" or remembrances of past experiences with games involving dice.

For example, on the first day of experimentation in classroom A (rolling two six-sided dice) the students tried to explain their predictions and outcomes:

Interviewer: Why do you get so many more fives? . . .
S1:   Well, it has always been my lucky number you know. . . .
I: S2, you said that you would get the most of 6. . . . Why did you think that? . . .
S2:   Lots of people roll 6s because I play lots of games at home, and I always roll 6s.

S1 thought he'd get fives more than any other outcome because fives were lucky to him. S2 recalled rolling a lot of sixes when playing games at home. Neither based their predictions or explanations on a model of chance.

Generalizing Over Past Experiences. After some experiences with seeing departures from predictions (i.e., residuals), most of the students attempted to change their prediction strategies so as to minimize residuals. The strategy that most of the children adopted was to generalize over a set of past experiences and change their predictions in such a way as to reflect that generalization. For instance, on the third day of investigation in classroom A, the children made predictions about the most and least frequent results when rolling two eight-sided dice. A number of the students modified their prediction strategies based on previous days' results. For example, some students noticed that with two six-sided dice, 6 occurred often. Moreover, 6 was the highest number on a single side of a die. Reasoning by analogy, they predicted that 8 would be the most frequent number with two eight-sided dice.

Basing Predictions on Sample Spaces. A minority of students came to understand residuals in terms of the relationship between outcomes and sample spaces. For example, in classroom B the teacher rolled two six-sided dice and graphed the results. The next day, students in her classroom replicated her experiment, first making predictions about their own graphs. In the following transcript, S9 justifies his prediction by making reference to the sample space and the uncertainty of individual outcomes:

I: Are your results looking like hers [the teacher's]? . . .

S3: I don't remember [what she got], but when we end it probably won't.

I: It probably won't when you're done? How come?

S3: Well, because we're shaking it, and we're probably going to shake something else . . .

I: Would everybody's be different?

S3: Well, it would be quite the same . . . it might be kind of the same.

I: How so would it be kind of the same?

S3: Well . . . 6, 7, and 8 would get kind of high . . .

I: On everybody's 6, 7, and 8 would get kind of high [occur more frequently than other outcomes]? How come?

S3: Because there's lots of chances [i.e., many different ways] of getting them.

S3 understood and accepted the role of chance in creating residuals (i.e., making everybody's graph different). He realized that his graph would probably end up looking different than his teacher's. He also understood, however, that due to the relationship between the sample space and the outcomes, everybody's graph would be "quite the same" in that "6, 7, and 8 would get kind of high . . . because there's lots of chances of getting them."

S3 and a few other students like him understood the role of the sample space and how it related to outcomes. For them, predictions were based on sample spaces, not on generalizations of past experiences. Although reasoning based on generalized past experiences is closer to model-based reasoning than reasoning based on lucky numbers, reasoning based on sample spaces, with an acceptance of residuals, essentially is conventional model-based reasoning in this domain.

In summary, when experimentation began the second-grade children did not in any way base their predictions on a model. Without a model, they had no way of accounting for the residuals that they observed. Most students took the presence of residuals as an indicator for the need to change their predictions. They did so by generalizing over past experiences. However, generalizing over past experiences is not the same thing as reasoning via a model of the domain. A few children were able to reason independently via a model of the domain based on the sample space. They accepted and expected random deviations from a predicted pattern of outcomes, but still based predictions and explanation on the sample space consistent with model-based reasoning.

MODEL-BASED REASONING ABOUT CHANCE IN THE FOURTH/FIFTH GRADE

We now turn to the performance of the second group of subjects—a single pair of fourth/fifth-grade students. These older subjects provide an example of how children's understanding of chance and uncertainty might develop. For a more detailed analysis of the performance of these older children, see Lehrer, Horvath, and Schauble (1994).

Distinguishing Between Certainty and Uncertainty

The fourth/fifth-grade students had no difficulty recognizing the uncertain nature of the spinners tasks they were given: predict, experiment, and explain the results of experiments involving one or two spinners. By most accounts (e.g., Byrnes & Beilin, 1987; Byrnes & Overton, 1986; Piaget & Inhelder, 1951/1975; Somerville, Hadkinson, & Greenberg, 1979), children of this age (i.e., 9–10 years old) are quite capable of recognizing situations of uncertainty and treating them appropriately, especially given the obvious randomizing nature of the spinner tasks (Metz, in preparation).

The Nature of an Experimental Trial

With one small exception, the fourth/fifth-grade children treated all spins of their spinners as comparable and combinable. In this sense, their understanding of the nature of the task and how it informed the consideration of the independent variables (i.e., those to be held constant) in the task seemed consistent with an understanding of the fundamental nature of the task—an (unbiased) investigation of uncertainty.

The one exception to this occurred when the students generated some results that did not meet with their expectations. At that time, and only at that time, the students questioned whether or not there were other features of the trials that might be accounting for the noticed residuals. They questioned whether or not the possible arch of the cardboard or the direction of geographic north might have some effect on the outcomes and, hence, should be held constant during trials.

S1: Well, less 2s. We didn't do that much. I think it's this side [pointing to the 2 and the 3] . . . This side is bigger [i.e., will have higher frequencies of outcomes] . . .

Investigator: So you think the numbers on this side of the spinner are . . .

S2:   Maybe it's because it's pointing to the north . . .

I: Maybe. Ok, so somehow you think that the ones on this side [pointing
      to the 2 and the 3] are more likely than the ones on this side [pointing
      to the 1 and the 4] now?

S1:   Yeah. Maybe the arch of the cardboard or something . . .

When the students noticed results that were not what they expected,
they, in essence, wondered if the mechanisms involved in the experiment
(the cardboard spinner) or other external forces (the direction of magnetic
north) were introducing any bias into the results. An introduction of bias
was unacceptable. As a whole, then, the fourth/fifth graders seemed to
understand the combinable nature of experimental trials and the need to
hold key task structures, (lack of) bias and uncertainty of outcomes, constant
in order to validly compare those trials.

Relationships Between Simple Events and Distributions

As discussed previously, the fourth/fifth-grade students clearly understood
the uncertain nature of individual (i.e., local) outcomes for the tasks in which
they were engaged. From the outset of the investigation, however, they
demonstrated that they thought that patterns of results could be predicted.
Initially they thought that each of the possible outcomes was equally likely.
For example, they predicted an equal value for each of the three possible
results (1–3) on the initial single spinner task and (initially) an equal value
for each of the five possible results (2–6) on the initial double spinner task.
After some experimentation with the task, however, they began to take the
patterns of results that they were generating into account for their future
predictions (e.g., they noted the "triangley" shape of the pattern of results
for the double spinner case and predicted "triangley"-shaped distributions
for future double spinner cases).

As a whole, the fourth/fifth graders demonstrated a clear understanding
that although local events are uncertain, patterns of events are predictable.
The patterns of events that they predicted changed throughout the experiment in response to the results they were generating; nevertheless, they
continued to expect and predict patterns of results. Like the second graders,
they tended to rely on previous experiences, not model-based ideals, as a
guide to the revision of predictions.

Structure of Events

During the early portions of experimentation the students made very little
mention of any aspect of the sample space and never based any of their
predictions or explanations on it, even after being explicitly asked to generate

the sample space for the task they were working on. In order to see if they could reason about these relationships, they were introduced to a method for systematically generating each permutation and to a notational system (a bar graph) for recording the number of permutations for each possible result—just as the second graders were. After only a short introduction (approximately 10 min) to the notational system that reified the sample space for them, the fourth/fifth-grade students were able to incorporate information about the sample space (e.g., the relation between relative frequencies of results in the sample space and expected relative frequencies of results in the outcome space) into their predictions about dice and spinners and did so on two transfer tasks (one using dice and one using spinners).

Although the fourth/fifth-grade students did not seem to originally understand the relationship between sample spaces and actual outcomes, with the support of the notational system provided to them they were able to incorporate it quickly and easily into their models of chance and uncertainty and to independently use that information. Further, unlike many of the second-grade children, when these forms of assistance were removed, the fourth/fifth-grade children were able to continue to reason about those relationships (i.e., the model retained those relationships even after the forms of assistance were removed).

Treatment of Residuals

During the first phase of experimentation, the fourth/fifth graders predicted that if they spun a single three-region spinner 25 times, each result would occur with equal frequency (as much as possible). After experimenting, however, they noticed that their results did not precisely match the predictions they had made (i.e., there were residuals). They quickly concluded that there would be some deviations between their predictions and their outcomes, that the match would not be "exact." They therefore began trying to account for the residuals they were seeing by consciously adding random variations to their predictions. After they had generated a few sets of results, they further noticed some patterns to their outcomes. When they noticed that "the middle numbers got more," they started to predict distributions with higher frequencies for central values.

Just as a portion of the second-grade children changed their predictions based on past residuals, so did the fourth/fifth graders. The fourth/fifth graders also understood both the role of the sample space in determining the basic distribution of results ("the middle numbers got more"), but also that chance variation would likely yield results that were slightly different from that distribution, just as the second grader S9 did.

In summary, the model of chance and uncertainty that the fourth-fifth grade children developed was reminiscent of some of the models that we saw in the second-grade children. The older children, however, needed much less assistance to reason about the uncertain phenomena they were investigating. They were also able to apply their models of those uncertain phenomena to new tasks much more easily than were the younger children. In this sense, the older children's models were much more powerful, applying to a wider range of tasks with less assistance.

MODEL-BASED REASONING ABOUT CHANCE IN ADULTS

The final group of subjects that we looked at was a pair of graduate students acquainted with conventional models of statistics. These subjects were chosen to represent an endpoint (of sorts) of the development of the concepts of chance and uncertainty. They treated all outcomes for the tasks they participated in (involving spinners) as uncertain. It was assumed, therefore, that they understood the concept of uncertainty for the spinners tasks that they participated in. They compared and combined all experimental outcomes. We took this as evidence that they understood the nature of an experimental trial and treated lack of bias and uncertainty of outcome as the only fundamental task structures. Their reasoning about the relationships between individual results and patterns of results was intimately tied to their understanding of how the structure of events related to outcomes. Throughout experimentation, they understood that local outcomes were highly uncertain whereas global patterns of outcomes were certain, more so with larger sample sizes. They expected residuals, recognizing that they were basing their predictions on a model of chance and uncertainty, that their model was based on an ideal and that deviations from that ideal (i.e., residuals) were to be expected.

In short, the adults seemed to reason about the tasks they were involved in much the same way as we would expect an "expert" modeler to. They understood the basic concepts involved in tasks of uncertainty (e.g., the nature of uncertainty, the nature of experimental trials) as well as the fundamental nature of modeling (i.e., a model represents a theoretically motivated abstraction that embodies certain key relationships of the phenomenon being modeled).

DISCUSSION

We have looked at three snapshots of the development of an understanding of chance and uncertainty: two classrooms of second-grade students, a pair of fourth/fifth-grade students, and a pair of statistically sophisticated adults. For each group, we analyzed their performance along five dimensions of a model of chance and uncertainty: (a) the distinction between certainty and uncertainty, (b) the nature of experimental trials, (c) the relationship between individual outcomes and patterns of outcomes, (d) the structure of events, and (e) the treatment of residuals.

The Growth of Model-Based Reasoning

Distinguishing Between Certainty and Uncertainty. We saw that even the second graders had a significant understanding of the certainty/uncertainty distinction for devices involving simple randomization. They clearly distinguished die-rolling tasks from other more determinate tasks. Although a number of students began experimentation believing that they could somehow influence outcomes when rolling dice, most quickly abandoned these ideas and viewed outcomes of die-rolling experiments as uncertain. Both older groups of subjects clearly understood the distinction between certainty and uncertainty and viewed the results of the tasks that they were engaged in as uncertain throughout the investigation.

The Nature of an Experimental Trial. The second graders, as a whole, did not understand the nature of experimental trials for the tasks they were engaged in using classical randomization devices as well as either of the older two groups did. The second graders were split as a class on whether or not they should view anything beyond uncertainty as a key structure of the task. Some, but not all, of the students also attended to issues of bias. In other words, they wanted to ensure that the results of each die toss were not influenced in any way that might make any one outcome (e.g., the outcome they predicted as most frequent) any more or less likely than another. Some of the students accepted bias in their outcomes as long as the results were still uncertain. The fourth/fifth graders and the adults both understood throughout the investigation that lack of bias and uncertainty of outcome (and nothing else) needed to be maintained between experimental events involving classical randomization devices.

Relationships Between Simple Events and Distributions. The second-grade children varied widely in their reasoning about the relationships between simple events (i.e., the local) and distributions of events (i.e., the global). All of them viewed the local as uncertain for tasks involving dice (e.g., the outcome of a single roll of a die). A number of them also thought the global (e.g., the distribution of results for rolling dice) was uncertain. Some, however, saw the global as predictable even though their predictions for what that global would look like differed. The fourth/fifth-grade students viewed individual elements (e.g., a single spinner turn) as unpredictable but distributions of results as predictable. The nature of the patterns that they predicted for the global, however, varied as their understanding of the sample space dimension developed. The adults, like the fourth/fifth graders, saw the local as unpredictable and the global as predictable throughout. Their understanding of why and how the global was predictable was based entirely on their understanding of the sample space dimension throughout the investigation. The influence of the groups' understanding along the sample space dimension on their understanding along the local/global dimension increased markedly from the second graders up through the adults.

Structure of Events. The second graders' understanding of the structure of events is also best characterized by diversity. Most of them failed to see any relationships between sample spaces and outcomes when the investigation began. After providing them with two forms of assistance, a bar graph notational system and the conversational support of the class, most of the second graders were able to understand the concept of a sample space and how it might relate to actual outcomes. When the forms of support were removed, however, few of them continued to reason about those relationships. At the beginning of experimentation, the fourth/fifth-grade students predicted equal frequencies for each outcome, ignoring information about the sample space. With some assistance from the teacher/instructor, however, and the support of a notational system (a bar graph for notating and reifying the sample space), they readily incorporated key ideas about the sample space into their reasoning and continued to do so even after the forms of assistance were removed. The adults understood throughout the investigation how the sample space related to outcomes (i.e., it was a representation of the ideal distribution of outcomes) and based every decision on those relationships.

Treatment of Residuals. There was a marked difference between the two older groups and the majority of the second graders in their treatment of residuals. The adults understood the nature of, and the reasons for, residuals in terms of a modeling process. They understood that the model of chance and uncertainty that they were employing was an idealization of the physical phenomena with which they were experimenting, the spinners. As an idealization, they realized that there would be differences between the ideal and the actual. Hence, residuals were a natural part of the experimentation process and were explainable by their model.

Throughout experimentation, the fourth/fifth-grade students based their predictions on a sample space of the domain. After noticing some residuals, their first response was to modify their predictions based on those past results and residuals. The students quickly came to accept residuals as an unavoidable consequence of experimentation in their domain. By the end of the investigation, however, their predictions were no longer based on generalizations of past experiences, but on a model of the domain as indicated by the sample space.

The majority of the second-grade children also quickly came to understand deviations from predictions (i.e., residuals) as an unavoidable consequence of experimentation in an uncertain environment. However, they handled residuals not by reasoning in terms of a model and explaining deviations in terms of a sample space, but merely by changing predictions so as to better fit generalizations over past experience.

Reprise

What have we gained by taking this modeling perspective in our study of the development of the concepts of chance and uncertainty? Statistics is all about modeling data. A model-based perspective, therefore, brings the teaching and learning of statistics in line with the nature of the domain, modeling. By considering in depth the dimensions that make up a model of chance and uncertainty, we are better able to understand and to assess the various models that different learners construct to understand uncertain phenomena. This serves as an appropriate starting point for the design of instruction, something best not left to chance.

REFERENCES

Accredelo, C., & O'Connor, J. (1991). On the difficulty of detecting cognitive uncertainty. *Human Development, 34*, 204–223.

Braine, M. D. S., & Rumain, B. (1983). Logical reasoning. In P. Mussen (Ed.), *Handbook of child psychology* (Vol. 3, pp. 263–340). New York: Wiley.

Byrnes, J. P., & Beilin, H. (1987). *The relationship between causal and logical thinking in children*. Paper presented at the biennial meeting of the Society for Research in Child Development, Baltimore, MD.

Byrnes, J. P., & Beilin, H. (1991). The cognitive basis of uncertainty. *Human Development, 34*, 189–203.

Byrnes, J. P., & Overton, W. F. (1986). Reasoning about certainty and uncertainty in concrete, causal, and propositional contexts. *Developmental Psychology, 22*(6), 793–799.

California Assessment Program. (1991). *A sampler of mathematics assessment*. Sacramento, CA: California Department of Education.

Cobb, P., Yackel, E., & Wood, T. (1993). Theoretical orientation. In Rethinking elementary school mathematics: Insights and issues. *Journal for Research in Mathematics Education*, Monograph No. 6, 21–32.

Davies, C. M. (1965). Development of the probability concept in children. *Child Development, 36*, 779–788.

Fennema, E., Franke, M. L., & Carpenter, T. P. (1993). Using children's mathematical knowledge in instruction. *American Educational Research Journal, 30*(3), 555–583.

Fischbein, E. (1975). *The intuitive sources of probabilistic thinking in children*. Boston: D. Reidel.

Fischbein, E., Nello, M. S., & Marino, M. S. (1991). Factors affecting probabilistic judgements in children and adolescents. *Educational Studies in Mathematics, 22,* 523–549.

Green, D. R. (1982). *Probability concepts in school pupils aged 11–16 years.* Reading, MA: Addison-Wesley.

Hawkins, A. S., & Kapadia, R. (1984). Children's conceptions of probability—A psychological and pedagogical review. *Educational Studies in Mathematics, 15,* 349–377.

Hestenes, D. (1992). Modeling games in the Newtonian world. *American Journal of Physics, 60,* 440–454.

Horobin, K., & Accredelo, C. (1989). The impact of probability judgements on reasoning about multiple possibilities. *Child Development, 60,* 183–200.

Kaput, J. J. (1991). Notations and representations as mediators of constructive processes. In E. von Glasersfeld (Ed.), *Radical constructivism in mathematics education* (pp. 53–74). Boston: Kluwer.

Kaput, J. J. (1992). Technology and mathematics education. In D. A. Gruows (Ed.), *Handbook of research on mathematics teaching and learning* (pp. 515–556). New York: Macmillan.

Kuzmak, S. D., & Gelman, R. (1986). Young children's understanding of random phenomena. *Child Development, 57,* 559–566.

Lehrer, R., Horvath, J. K., & Schauble, L. (1994). Developing model-based reasoning. *Interactive Learning Environments, 4*(3), 218–232.

Lehrer, R., Jacobson, C., Thoyre, G., Gance, S., Horvath, J., Kemeny, V. & Koehler, M. (in press). Mathematizing space in the primary grades. In R. Lehrer & D. Chazan (Eds.), *Designing learning environments for developing understanding of space and geometry.* Hillsdale, NJ: Lawrence Erlbaum Associates.

Moore, C. Bryant, D., & Furrow, D. (1989). Mental terms and the development of certainty. *Child Development, 60,* 167–171.

National Council of Teachers of Mathematics. (1989). *Curriculum and evaluation standards for school mathematics.* Reston, VA: Author.

Newell, A. (1980). Physical symbol systems. *Cognitive Science, 4*(2), 135–183.

Newell, A., & Simon, H. (1981). Computer science as empirical inquiry: Symbols as search. In J. Haugeland (Ed.), *Mind design: Philosophy, psychology, artificial intelligence* (pp. 35–66). Cambridge, MA: MIT Press.

Piaget, J., & Inhelder, B. (1975). *The origin of the idea of chance in children.* New York: W. W. Norton. (Original work published 1951)

Resnick, M. (1994). *Turtles, termites and traffic jams.* Cambridge, MA: MIT Press.

Shaughnessy, J. M. (1992). Research in probability and statistics: Reflections and directions. In D. Gruows (Ed.), *Handbook for research in mathematics education* (pp. 465–494). New York: Macmillan.

Somerville, S. C., Hadkinson, B. A., & Greenberg, C. (1979). Two levels of inferential behavior in young children. *Child Development, 50,* 119–131.

Sophian, C., & Somerville, S. C. (1988). Early developments in logical reasoning: Considering alternative possibilities. *Cognitive Development, 3,* 183–222.

Stewart, I. (1989). *Does God play dice?: The mathematics of chaos.* New York: Basil Blackwell.

Stigler, S. (1986). *The history of statistics: The measurement of uncertainty before 1900.* Cambridge, MA: Harvard University Press.

Vygotsky, L. (1978). *Mind in society.* Cambridge, MA: Harvard University Press.

CHAPTER SIX

Emergent Ideas of Chance and Probability in Primary-Grade Children

Kathleen E. Metz
University of California, Riverside

The goals of this chapter stem from the conjunction of two different recommendations of the mathematics reform movement today. The National Council of Teachers of Mathematics (NCTM) contended that children's mathematics instruction should be grounded in their intuitions, "buil[t] upon the concepts and principles that children already possess" (NCTM, 1989b, p. 33). Both NCTM's influential *Curriculum and Evaluation Standards for School Mathematics* and the American Statistical Association–NCTM Joint Committee on the Curriculum in Statistics and Probability (1985) contended that statistics and probability belong in the curriculum beginning at the primary-grade level. If primary-grade teachers are to ground their mathematics instruction in children's ideas, and if they are to include statistics and probability in their curriculum, then they need to know the intuitions relevant to initial statistics and probability that primary-grade children bring to school.

However, a top-level analysis of the research literature bearing on children's thinking in this area reveals that this research literature is rife with contradictions concerning what children of different ages understand of key ideas in this domain. For example, Piaget and Inhelder (1975) concluded that "chance" first enters into the child's thinking around 7 years of age, as the differentiation of the predictable from the unpredictable, the determined event from the event that is undetermined and indeterminable. Kuzmak and Gelman (1986) reported that by age 5, or possibly 4, children understand chance and randomness, in the sense that they can differentiate random from deterministic phenomena. Fischbein (1975) argued that the intuition of randomness is manifested by the preschooler, albeit under limited conditions. At the other extreme, Green (1988) concluded that 11- to 16-year-olds' understanding of randomness is at best fragile.

The literature examining children's understanding of probability is similarly inconsistent. Yost, Siegal, and Andrews (1962) reported that even 4-year-olds have some understanding of probability. Falk, Falk, and Levin (1980) concluded that the first grader can understand probability, at least on some intuitive level. Fischbein (1975) reported that, with instruction, probability is accessible to the 9- or 10-year-old. However, Piaget and Inhelder (1975) concluded that probability emerges in adolescence, after the requisite proportionality and combinatorial schemes enter into the children's repertoire.

These contradictions in the literature might not pose such a problem if teaching-as-telling was still regarded as a adequate means for teaching mathematics. However, if we view mathematics learning in the classroom as a process of the children's active elaboration of their ideas, mediated and supported by the teacher's scaffolding (Cobb, Yackel, & Wood, 1993), than knowledge of the children's ideas in the domain becomes crucial. A research base that can inform teachers of the key ideas that children bring to instruction is particularly important in a domain as complex as statistics and probability, in which most teachers themselves have little training.

This chapter examines the power and challenge of successfully incorporating statistics and probability into children's mathematics curriculum, from the perspective of instruction grounded in children' ideas. It examines the power of the epistemological messages that incorporation of this curricular strand can support, epistemological messages concerning both the utility of mathematics as a interpretative tool and of a world view encompassing uncertainty and ambiguity. It subsequently examines the challenge of incorporating statistics and probability in the primary grade curriculum, in light of the inherent difficulty of the core constructs and the current state of the literature. Then it briefly reviews the instructional objectives of different countries that emphasize statistics and probability at this level and, on this basis, identifies five key ideas: relative magnitude, part /whole relations, incertitude and indeterminacy, the likelihood of an event, and expected distributions of outcomes. The developmental and mathematical cognition literatures are examined to identify the aspects of these ideas that children understand or fail to grasp, from the perspective that primary-grade teachers can build on and further scaffold these intuitions in their statistics and probability activities. The chapter ends with an examination of more general issues in the successful incorporation of a statistics and probability curricular strand at the primary grade level.

THE POWER OF INCORPORATING STATISTICS
AND PROBABILITY IN PRIMARY-GRADE MATHEMATICS

Epistemological Rationale

The NCTM and other corners of the mathematics education reform movement have voiced strong concern that the curriculum more closely reflect the nature of thinking in the discipline of mathematics. Incorporation of a statistics and probability curricular strand is strategic from this epistemological perspective. This section considers both how a statistics and probability focus at this age level can support a view of mathematics as a tool of interpretation and sense-making, useful in solving problems across other domains, as it also supports a view of how to reason in a nondeterministic world.

An important consideration in the design of mathematics curriculum concerns the idea about mathematics and the nature of mathematics activity that we convey to students. Children's mathematics curricula has been strongly criticized from this perspective. In the United States, arithmetic taught through drill and practice has traditionally dominated young children's mathematics curricula. As the NCTM noted, this approach results in children beginning "to lose their belief that learning mathematics is a sense-making activity" (NCTM, 1989a, p. 15). In this same vein, the National Assessment of Educational Progress (Dossey, Mullis, Linquist, & Chambers, 1988) indicated that many children have a very limited view of mathematics and mathematical activity. For example, half of the large sample of seventh-grade students agreed with the statement, "Learning mathematics is mostly memorizing" (p. 102). In a similar vein, the California Mathematics Framework noted that "Too many students have come to view mathematics as a series of recipes to be memorized, with the goal of calculating the one right answer to each problem. The overall structure of mathematics and its relationship to the real world are not apparent to them" (California State Department of Education, 1985, p. 12) In short, the arithmetic drill and practice model of children's mathematics supports an impoverished set of connections between mathematics and other domains or spheres of application.

If we aim to encourage children to value mathematics as a powerful tool of interpretation and sense-making, then a vision of mathematics that emphasizes external connections seems most powerful. Lynn Steen's vision of mathematics as the "science of patterns" beautifully captures this view:

> Mathematics is the science of patterns. The mathematician seeks patterns in number, space, in science, in computers, and in imagination. Mathematical theories explain the relations among the patterns. Applications of mathematics

use these patterns to "explain" and predict natural phenomena that fit the patterns. (Steen, 1988, p. 616)

Integration of statistics and probability into children's mathematics curriculum can support a vision of mathematics as a tool of sense-making and interpretation across the curriculum, of mathematics as the "science of patterns."

Incorporation of statistics and probability can also support another epistemological agenda, namely, a view of the world as nondeterministic with uncertainty and ambiguity. Statistician and statistician educator David Moore (1990) argued that "children who begin their education with spelling and multiplication expect the world to be deterministic, they learn quickly to expect one answer to be right and the others wrong, at least when the answers take numerical form." Fischbein (1975) argued that most schools fit this deterministic mode and actually have a detrimental affect on children's grasp of the core randomness construct. A curricular strand in statistics and probability can potentially support children's conceptualization of the world in nondeterministic terms, as it begins to scaffold their mathematical strategies for coping with the uncertainty.

Science Curriculum Rationale

Children's study of statistics and probability, in conjunction with their engagement in data analysis, can empower the possibilities of children's science curricula. Contemporary science educators are also concerned with the messages in the curriculum about the nature of science and what it means to do science (Burbules & Linn, 1991). Empirical investigations, including the analysis of data, are fundamental to an adequate vision of science and the inquiry process. In this vein, the American Association for the Advancement of Science contended that:

> From the very first day in school, students should be actively engaged in learning to view their world scientifically. That means encouraging them to ask questions about nature and to seek answers, collect things, count and measure things, make qualitative observations, organize collections and observations, discuss findings, etc. (American Association for the Advancement of Science, 1993, p. 6)

Incorporation of statistics and probability in children's mathematics curriculum can enhance the crucial empirical component of the children's science curriculum and thus strengthen the vision the science curriculum can convey to children about the nature of science and scientific activity.

Ironically, although many corners of the mathematics education reform movement view primary-grade children as potential investigators, capable of collecting, representing, and interpreting data (National Council of Teachers of Mathematics, 1989a; Romberg, Allison, Clarke, Clarke, & Spence, 1991), a common trend in science curricula (Rosier & Keeves, 1991) reveals a more modest view of children's capacities. In this approach, science curricula are frequently framed in terms of "developmentally appropriate" science process objectives. Such processes as observation, measurement, categorization, and seriation are emphasized at the primary-grade level. Empirical investigations are relegated to higher grades, on the assumption that planning of experiments and data analysis are beyond the reach of young children. However, a close analysis of the broad base of relevant developmental and nondevelopmental research indicates that these purported limitations on young children's scientific inquiry are not supported by the literature (Metz, 1995). Rather the vision of the child as empirical investigator, explicit in the mathematics reform literature, has considerable support.

In summary, instruction in statistics and probability in the primary-grade mathematics curricula can empower children's data analysis and consequently enhance the power of their scientific inquiry. Indeed, incorporation of this curricular strand appears at least as important to the reform of children's science curricula as it is to the reform of their mathematics curricula.

THE CHALLENGE OF INCORPORATING CHANCE AND PROBABILITY IN PRIMARY-GRADES MATHEMATICS

Despite compelling reasons to include statistics and probability in the primary-grades curriculum, there exist significant challenges in its successful implementation. These difficulties stem from both the complexity of the domain and the current state of the research and theoretical literatures.

The statistical literature conceptualizes randomness and probability as aspects of the same construct. As explained by statistician David Moore:

> Phenomena having uncertain individual outcomes but a regular pattern of outcomes in many repetitions are called *random*. "Random" is not a synonym for "haphazard," but a description of a kind of order different from the deterministic one that is popularly associated with science and mathematics. Probability is the branch of mathematics that describes randomness. (Moore, 1990, p. 98)

Within this frame of thinking, Steinbring (1991) argued that one cannot understand chance without a grasp of probability and, conversely, one cannot understand probability without a grasp of chance.

Although the expert's conceptualization of randomness and probability integrates these ideas into the same construct, the cognitive developmental literature has frequently separated the two. Thus some studies have focused on the emergence of chance, as indicated by the child's attribution of uncertainty and unpredictability. Other studies have focused on the patterns emerging across many repetitions of an event—frequently with little attention to whether or not the child understands the uncertainty associated with the patterns.

Another distinction that appears to be a source of confusion in the children's cognitive literature is that fact that there are multiple types of probability (classicist probability, frequentist probability, subjective probability), most which can be conceptualized either in terms of expected frequencies or in terms of the likelihood of a single event. However, the child cognition literature typically does not make these distinctions, thus considering as a singular challenge and accomplishment adequate problem solving across cognitively distinct scenarios.

Although there is widespread disagreement concerning what children actually understand of chance and probability, the cognitive research literature at this age and older documents the considerable difficulty that reasoning under conditions of uncertainty poses to all ages. The inherent difficulty of the domain, even at the adult level, highlights the significant challenge of effective instruction at the primary-grade level.

INSTRUCTIONAL OBJECTIVES IN PRIMARY-GRADES STATISTICS AND PROBABILITY, CRITICAL IDEAS

For the purpose of identifying concepts that are relevant to primary-grade instruction in statistics and probability, we consider the instructional objectives from three countries that include a relatively strong emphasis in this sphere: the United Kingdom, Australia, and the United States (Romberg et al., 1991). Examination of their respective recommendations for appropriate instructional objectives and activities in primary grade statistics and probability reflect considerable overlap (Table 6.1).

All three countries recommend structuring children's reflection on chance, albeit with different emphases and different implied instructional approaches. All three countries assume that children should conduct simple investigations. Furthermore, they assume that children should be engaged in all phases of this process, from question formulation to data collection, data structuring, and data interpretation. Although the United Kingdom and Australian perspectives appear to place a greater emphasis on probability, analysis of the details and illustrations in the U.S. document, *Curriculum and Evaluation Standards for School Mathematics,* reveals the assumption

TABLE 6.1
Comparison of Recommended Activities for
Primary Grade Statistics and Probability

Country	Instructional Objectives: K–4 Statistics & Probability
Australia	"Experiences should be provided which enable children to: —Use, with clarity, everyday language associated with chance events —Describe possible outcomes for familiar-random and one-stage experiments —Place outcomes for familiar events and one-stage experiments in order, from those least likely to happen to those most likely to happen —Frame simple questions about themselves, families, and friends, and collect, sort, and organize information in simple ways in order to answer these questions —Represent and interpret information to answer simple questions about themselves, friends, and family" (quoted from Romberg et al., 1991, Appendix, p. 11)
United Kingdom	"Sort a set of objects, describing criteria chosen Recognizes that outcomes can vary Interpret data that has been collected by pupils Recognize that there is a degree of uncertainty about the outcomes of some events but that others are certain or impossible Access information in a simple database Construct and interpret statistical diagrams Use appropriate language to justify decisions when placing events in order of likelihood Interrogate and interpret data in a computer database Conduct a survey on an issue of (their) choice and communicate results Use the mean and range of a set of data Estimate and justify the probability of an event" (quoted from Romberg et al., 1991, Appendix, p. 12)
United States	"The mathematics curriculum should include experiences with data analysis and probability so that students can— • collect, organize, and describe data • construct, read, and interpret displays of data • formulate and solve problems that involve collecting and analyzing data • explore concepts of chance" (quoted from NCTM, 1989a, p. 54)

that the children should explore ideas of probability as well. Underlying all three approaches is the view that we should begin to develop children's knowledge of statistics and probability at the primary grades through their active engagement in data analysis of meaning to them and their guided reflection on key statistical and probabilistic ideas within this context.

Analysis of these curricular objectives reveals a number of fundamental ideas. This chapter focuses on five. First, analyses of data, even on the most rudimentary and intuitive level, involve some kind of comparison of quantities or frequencies, that is, some form of relative magnitude. Second, part–

whole relations are fundamental to understanding the meaning of probabilities. For example, relative probabilities involve the notion of the sum
of probabilities as the whole; the probabilities of all possible outcomes total
1. Conceptualizing degree of likelihood on a qualitative and continuous
parameter, from 0 as impossible to 1 as certain, implies at least an implicit
comparison of the weight of the outcome under consideration with the
weight of certitude. Third, uncertainty and indeterminacy, including the idea
that there is more than one possible outcome for a given event, are also
fundamental to reasoning within the domain of statistics and probability.
The last two ideas concern different ways of conceptualizing probabilities:
probabilities as the likelihood of a given event, and probabilities as expected
distributions of outcomes.

ANALYSIS OF CHILDREN'S INTUITIONS

The following subsections examine the research literature to identify what
aspects of each of the five ideas primary-grade children appear to have
constructed independently. Almost all of the research concerning children's
statistical and probability cognition has been conducted within the developmental paradigm. Therefore there has been little research to date at this
age level investigating what children can learn, given appropriate instruction.
Thus the research examined here can function, not as model of the potential
of children's understanding with instruction, but as ideas on which they can
base instruction.

Relative Magnitude

Analysis of probabilities demands some understanding of relative magnitudes. For example, the ability to distinguish between rarity and high frequency of a particular kind of element within a collection is fundamental
to reasoning about probabilities. The mathematical cognition literature indicates that children come to kindergarten with the rudiments of this knowledge and that it is further elaborated and extended across the primary-grade
years.
    Knowledge of relative magnitude implies an ability to quantify classes of
elements and then to compare the number of elements in the different
classes, at least on an approximate level. According to the literature, children
have multiple means of assessing quantity even before they enter kindergarten. Gelman and Gellistal (1978) reported that although their counting
may reflect errors, preschoolers understand the principles underlying counting. Klahr and Wallace (1976) documented the ability of preschoolers to
subitize, defined as assessing a quantity through direct perception as op-

posed to counting. Resnick, Lesgold, and Bill (1991) used the term *proto-quantitive reasoning* to refer to the nonanalytical, direct perceptual strategy of quantification they attribute to the preschooler.

Various corners of the research literature also document that ability of children entering kindergarten to implement comparisons of quantities. Again the literature points to a range of strategies for this purpose. Resnick et al. (1991) identified what they termed the *protoquantitative compare* scheme, where children make comparisons of relative magnitude on the basis of their perceptually derived quantifications of the two sets. Siegler and Robinson (1982) reported that by 5 years of age children can accurately identify which number is more when simply given the pair of numbers in oral form. These authors assumed the ability is based on a number-line mental representation, given the findings that pairs of numbers with smaller differences take the children longer to compute.

Alongside these considerable strengths, 5-year-olds' thinking reflects limitations in the assessment of relative quantities. Not surprisingly, higher numbers can pose problems for young children (Siegler & Robinson, 1982). Furthermore, spatial aspects of the collections (e.g., where one collection is more spread out than another) can still mislead the children's judgments of comparative quantities (Piaget & Szeminska, 1965; Siegler, 1981) or may even lead them to question the results of their comparative counts (Piaget & Szeminska, 1965). These limitations largely disappear across the primary-grade years.

Part–Whole Relations

Understanding of part–whole relations underlies the ability to reason about probabilities. The research literature indicates that 5-year-old children bring important intuitions about the relation of part and whole with them to school. There is also dramatic development of the idea across the primary-grade years.

Resnick et al. (1991) argued that children enter school with the proto-quantitative part–whole scheme, namely, knowledge about how parts and wholes come apart and go together in the real world and relations between them. Thus, they explained, children know that a whole cake is larger than any of its pieces. Aspects of the Carpenter, Ansell, Franke, Fennema, and Weisbeck (1993) finding of kindergartners' remarkable competence in modeling actions and relations embedded in story problems involving addition, subtraction, multiplication, and division appear to reflect a rich understanding of numerical parts and wholes and the flexible use of this knowledge in the process of mathematical modeling. Saxe's (1991) descriptions of unschooled child street-sellers' transactions reflect the part–whole scheme, including its relatively complex and probability-relevant form of comparative ratios. Manifestations of the schema in this context support its intuitive origins and the potential power of the child's modeling abilities based on elaborations of this scheme.

Information-processing analyses of children engaged in different aspects of the primary grade mathematics curriculum indicate the centrality of the part–whole schema. For example, it underlies children's solving of many forms of story problems (Riley, Greeno, & Heller, 1983). Similarly, children frequently rely on the part–whole schema in their invention of algorithms (Resnick, 1983, 1986). This crucial schema underlies many aspects of the mathematics curricula, including the Base 10 number system, the operations of addition, subtraction, division and multiplication, fractions, and so on.

In short, the part–whole schema is evident in some forms by kindergarten level and more elaborated forms later during the primary-grade years. In terms of the statistics and probability curricular strand, the challenge involves helping children extend their ideas of part–whole relations into conceptualizations of probability and supporting their use of these ideas in the modeling of probabilistic situations.

Incertitude and Indeterminacy

Any consideration of statistics or probability, even at an intuitive and qualitative level, demands some understanding of the existence of uncertainty and the possibility of multiple outcomes for a given event. There is converging evidence that by 5 years of age children grasp this fundamental idea. The research literature also identifies many aspects of the construct of uncertainty that emerge during the primary grade years or much later. Indeed, many aspects of uncertainty pose considerable difficulty to adults (Kahneman, Slovic, & Tversky, 1982; Konold, 1991; Vallone & Tversky, 1985).

Different criteria corresponding with different conceptualizations for what it means to understand chance appear to constitute a major factor underlying the conflicting ages reported in the literature. Examination of these different criteria and the age at which they are supposedly met provides a window on the process of conceptual elaboration. A relatively rudimentary form, appearing around 4 or 5 years of age, is the distinction between phenomena with certain and predictable outcomes versus phenomena with uncertain and unpredictable outcomes. Kuzmak and Gelman (1986) argued that their preschool subjects demonstrated this understanding by differential responding to the question, "Do you *know* which color is going to come out next? Yes or No?" in reference to a deterministic apparatus (a plastic marble tube with a hole at one end) versus a nondeterministic apparatus (a wire steel case from which a colored marble would occasionally fall). By age 5, they reported, most children were able to give "appropriate explanations" of why the latter apparatus was unpredictable (operationalized as the child's reference to the fact that the marbles in the cage were mixed up, or noting different marbles moved across the opening, or that one couldn't see how the marbles were moving). The problem solving that Kuzmak and Gelman interpreted as evidence of children's understanding of random phenomena appears relatively elementary, albeit a core aspect of this complex construct.

Fischbein (1975) reported that preschoolers are quite adept at differentiating determined from chance situations under certain circumstances. Fischbein identified situations with two equiprobable outcomes as the easiest and reported, not surprisingly, that preschoolers cannot manage situations that demand analysis of combinations and permutations. Fischbein, however, was careful not to attribute a full understanding of incertitude to these young children. He contended that preschoolers manifest a "relative lack of differentiation between chance and caprice and between the arbitrary and the possible" (1975, p. 70) and, more generally, their knowledge is intuitive, not conceptual (i.e., without any instructional intervention, they have less awareness of the idea than older children who grasp the construct at the conceptual level).

Piaget and Inhelder (1975) would not consider any of the constructions identified by Kuzmak and Gelman or Fischbein to constitute an understanding of chance. According to Piaget and Inhelder, understanding of chance includes a clear differentiation of the necessary from the possible. It also presumes that the children have developed beyond a propensity to impose order or deterministic causality in all phenomena, having realized that many phenomena (such as path of the fall of a leaf) have no underlying reason or straightforward causality. More subtle and, Piaget and Inhelder claimed, fundamental to the concept of randomness, the children have also developed some understanding of both the interaction and independence of causal events.

The appearance of inconsistency across the preschool and school-age research is heightened by the fact that Piaget and the literature he inspired have frequently framed descriptions of thinking at the preoperational level as a foil to the competence of subsequent years. In response, an underlying agenda of much of the preschool research has been to rebut the theoretical position of a fundamental shift in adequacy of cognition from preschool to school-age children (Bullock, 1985). Neither Piaget's theoretical or methodological approach demanded an analysis of the aspects of knowledge of chance that younger children did possess. These differences in perspectives and the influence they have on how competent or incompetent the children appear is dramatically reflected in Fischbein's identification of positive aspects of younger children's intuitions from the exact same excerpts that Piaget and Inhelder used to illustrate their conceptual incompetence:

"Mon (4.10). 'Where will the marble go?' 'Perhaps there, or there' (pointing to the two possible exits). (The trial is run, and the marble rolls toward the right-hand exit.) 'And the next?' '*There*' (Trial: left-hand exit.) What about the rest of the marbles?' *'I don't know . . .'* " (Piaget and Inhelder, 1975, p. 44). (The experimental apparatus consists of five rectangular boxes placed on a slope, with a number of exits at the bottom.)

Roulette experiment: "Mon. (4.11). 'Can we tell where it will stop?' *'No, because if we say it will stop at blue, and then it goes past blue, we won't know.'* " (*ibid.*, p. 74) (cited in Fischbein, 1975, pp. 70–71)

Fischbein pointed to evidence of children's intuition of incertitude in these examples, in the sense that they emphasize the possibility of multiple outcomes. Piaget and Inhelder focused on their failure to conceptualize the idea of a distribution formed by the movement of all the marbles.

These different corners of the developmental literature, combined with research and theory of the adult judgment and decision-making literature, make clear the many levels of complexity involved in the notion of incertitude. The research indicates that 5-year-old children bring to school a grasp of incertitude in the sense of being able to consider what situations produce deterministic versus nondeterministic results and of correctly making the distinction in very simple situations. Seven- and 8-year-olds appear to have transformed the idea of randomness into a more explicit and refined concept, more clearly differentiating chance from caprice, the arbitrary from the possible, the possible from the necessary. These distinctions still leave open many fundamental issues, such as being able to think about the patterns that emerge and the variability that arise with the patterns, the source of the uncertainty, and the demarcation of phenomena involving an aspect of chance from those that are inherently deterministic.

The Likelihood of an Event

We can conceptualize probabilities in the singular or distributional mode. Conceptions of the likelihood of an event are based on the simpler singular mode. The idea of the likelihood of a single event is important in the conceptualization of probabilities and prediction of uncertain events. Subjective probabilities and both frequentist and classicist types of objective probabilities can be used to express degrees of likelihood about single events or distributions.

Subjective probabilities are framed in terms of an individual's degree of confidence or belief that a particular event will occur, for example, as the Australian document suggested, having children organize events in a continuum of their increasing confidence in the occurrence of the event from impossible to certain (e.g., from *It won't get dark tonight* to *When I get home today, my mom will be there*, etc.) Classicist probabilities are derived from analysis of the symmetries of a chance-generating device, such as spinners or dice. Frequentist probabilities are derived from trends that emerge over the course of empirical observations, under the assumption that the expected value associated with the targeted outcome will be approximated by its relative frequency across a large number of repetitions of the phenomena. The literature examined next indicates that primary-grade children grasp fundamental aspects of the likelihood-of-an-event construct in ways reflecting each of these probability forms. As with ideas of incertitude, fundamental aspects are also missing from the children's constructs.

Likelihood of an Event: Subjective Probability. The research suggests that key ideas of subjective probabilities begin to emerge prior to 5 years of age. In an intriguing study within the preschool "theory of mind" research literature, Moore, Pure, and Farrow (1990) found that by 4 years of age children understand that belief can be held at different levels of confidence. Furthermore, 4-year-olds can successfully use speaker information about levels of confidence and belief (as expressed in such terms as possibly and probably) as clues to identify the location of hidden objects. Moore et al. concluded: "Children who understand the representational nature of mind are able not only to recognize the existence of false beliefs in others, and changes in their own beliefs, but also to recognize that others' beliefs may be held with more or less certainty" (1990, p. 729).

Other research addressed the elaboration of subjective probabilities that subsequently emerge. Using more complex linguistic variations and a more rigorous task to perform with the information, Bassano (1985) reported dramatic development in discriminations of degrees of belief from 6 to 8 years of age. She concluded that various gradations of subjective probabilities are clearly differentiated and interpreted around 8 years of age. Huber and Huber (1987) reported remarkable competence in comparisons of subjective probabilities, which they concluded emerge around 5 years of age. These researchers identified six principles underlying children's judgments of subjective probability, including:

> If event A is more probable than event B, and B is more probable than C, then event A is also more probable than event C. . . .
> If event B is more probable than event C, then the combined event that A occurs or B occurs is more probable than the combined event than A occurs or C occurs, and *vice versa*. . . .
> If event A is more probable than event B, then the complement of event B is more probable than the complement of event A. (Huber & Huber, 1987, p. 306)

Using the relatively weak criterion of correct judgments, the researchers reported that "even young children apply this concept remarkably well, and no difference could be found between the age groups from 5 years to 17 years or older" (p. 313).

In short, the small extant developmental literature addressing the likelihood of an event in subjectivist terms reveals remarkable understandings on the part of primary grade children. These findings support the Hawkins and Kapadia (1984) contention that subjective probabilities may be a strategic type of probability on which to ground instruction.

Likelihood of an Event: Classicist Probability. The research indicates that young children have rudiments of classicist probability, in the sense that they infer information from the structure of chance-generating devices to construct anticipations. However, there is little evidence that they interpret this information in terms of expected distributions, but rather as the likelihood of a given event. Integrating such expectations with a sense of uncertainty and chance poses considerable difficulty, particularly at the lower end of the primary-grades age span.

For example, Falk et al. (1980) presented children with lottery-like tasks, where they were offered a prize if a spinner or top they selected from a given pair landed on their color. Falk et al. used the criterion of choice of device with the higher expected payoff as a sufficient indicator of probabilistic reasoning, even when the child was unable to produce any kind of verbal explanation. On this basis, they report that even the 5-year-olds' choices were better than chance, thus indicating an early grasp of probability. Nevertheless, they also posed as a possibility that their young subjects' correct choices may have stemmed more from perceptual judgments than an analytical intellectual frame. In other words, the subjects could have identified the device with the higher payoff color simply on the basis of which spinner or top had more of the targeted color, thus never considering the uncertainty and probability involved in the task.

Metz (in preparation) aimed to get at subjects' attribution of meanings underlying their problem solving in potentially classicist probability situations, through close analysis of their problem solving as recorded on videotapes. Given the substantial challenge that thinking probabilistically poses to adults as well as children, Metz compared cohorts of kindergartners, third graders, and university undergraduates. Each subject analyzed the effects of different spinners, where progress was determined by whether or not the spinner landed on your color during your turn. The spinners Potential manifestations of probabilistic reasoning included such situations as choice of spinner and rationale, anticipated effect of playing with a different spinner, projections concerning game outcome and where on the game board the anticipated loser will be when the game ends, implicit theories about spinner function, and reflections on what would happen if the game was repeated or a different spinner was substituted. Analyses revealed a complex pattern of improvement across the age span (macrodevelopment) and learning on task (microdevelopment).

Children's problem solving revealed that most of them had difficulty in integrating ideas of expectation, as derived from analysis of the spinner visual symmetries, with ideas of uncertainty. For example:

> The experimenter asks eight and a half year-old Mary to compare possible effects on the game outcome of playing with two different spinners; either a spinner with two evenly divided sectors, one red and one yellow versus a spinner with red and yellow in a ratio of 3:1.
>
> "Do you think it will make any difference which spinner we play with?" "Yes," responds Mary [touching the 1:1 spinner], "'Cause half of yellow and half of red." The experimenter probes, "And what will that mean in terms of the game? What difference will it make?" Mary explains, "'Cause maybe sometimes you turn it, it'll stop on the red and sometimes you turn it it'll stop on the yellow." "When you turn this [the 3:1 spinner] does it sometimes stop on the yellow and sometimes stop on the red?" Mary is asked. She shakes her head. "No!" When asked to predict what square the experimenter's yellow chip will beat at the end of the game, Mary exclaims, "There!" as she points to the starting point.

At this point in her protocol, Mary assumes that the player using the yellow chip will not advance a single step. Thus, for her the uneven colors in the spinner support not probabilistic but deterministic predictions.

Considerable development from kindergarten to third grade was manifested in this regard. Although integration of chance and probability was rare at the kindergarten level, it was eventually achieved by over half of the third graders. Other difficulties, manifested almost exclusively by the kindergartners, included the basic idea that it is not just the inclusion of each of the colors on the spinner that matters for what color it lands on, but also the color balance in the spinner; and the idea that the device is a chance-generating device, not controllable by expertise stemming from extended practice. Furthermore, in contrast to the undergraduates, where device-based expectations were manifested in the children's thinking, these expectations took the form of the likelihood of a given event, never the form of distributions.

Likelihood of an Event: Frequentist Probability. Although the developmental research literature does not usually identify children's thinking in these terms, close analysis of this literature indicates that children can spontaneously interpret frequentist probability in terms of the likelihood of a given event. For example, Acredolo, O'Connor, Banks, and Horobin (1989) designed their study to challenge Piaget and Inhelder's (1975) purported conclusion that, prior to adolescence and formal operational thought, children could only assess relative odds when the size of the sets under comparison was held constant. (Actually Piaget and Inhelder contended children could succeed when either the size of the sets or the number of favorable cases was held constant.) Acredolo et al. argued that Piaget's choice of methodology for studying probability knowledge (defined as children's comparison of the likelihood—less, equal, or more—of two outcomes) underestimated children's knowledge and proposed to replace it with an application of Anderson's (1980) functional measurement. Acredolo et al. stated that functional measurement assesses "whether children can integrate information in a way that mirrors the true algebraic relations between dimensions" (Acredolo et al., 1989, p. 934), as manifested in their placements of estimates on a physical sliding scale. In their translation of the functional measurement methodology to probability, they explained:

> We may ask children to estimate the likelihood of each of a series of discrete events on the basis of information about the number of potential outcomes and the total number of possible outcomes involved in each (i.e., the numerator and denominator in the probability calculation. (Acredolo et al., 1989, p. 934)

For example, one of Acredolo's experimental procedures presented the subject with a series of 12 computer-displayed images of a planter filled with pots of plants. From one image to the next, the researchers varied the total number of pots and the number of pots that had a spider. The subject is asked to assess the likelihood that a randomly bouncing bug would land on a plant with a spider, by selecting a placement on the scale marked at one end with a frowning bug and the other end with a smiling bug. The task is framed in terms of helping the bug. Their subjects, including cohorts of first, second, third, fourth, and fifth graders, all performed well. Indeed, the authors reported no significant age differences in competence of performance. Acredolo et al. interpreted their study as indicating that "even young concrete operational children can utilize denominator as well as numerator information when estimating probabilities." They argued that this knowledge is not revealed in the choice paradigm, due to the fact it somehow fosters nonoptimal strategies (perhaps, the researchers suggested, belief that comparison of numerators is an adequate basis on which to base assessments of comparative probabilities).

From this author's perspective, the study by Acredolo and his colleagues documents first graders' ability to competently assess the likelihood of a given event (e.g., that the bug will land on a plant with a spider), given the opportunity to express their assessment of probability on a sliding scale. It appears that two significant simplifications have been introduced here: the substitution of the singular mode for the distributional mode, and the number-line-like model by which children can express gradations of probability in qualitative form. A third factor that may also have functioned as a simplification was the situating of the probability into a kind of scenario they may have experienced.

The literature's various conceptualizations of what children grasp of probabilities and the strategies they employ reflect key methodological and theoretical issues, which Howell and Burnett (1978) addressed in their seminal paper examining our analysis of the cognition of uncertainty. Speaking from the perspective of the adult judgment and decision-making literature, Howell et al. emphasized the importance of going beyond analysis of correct or incorrect judgments. They argued that we need to closely analyze characteristics of the chance event under consideration (including seeming demands of the situation and how the subject has conceptualized the event), the cognitive process underlying the subjective probability judgment (including possible simplification rules), and task characteristics that may affect subject's choice of cognitive process (such as the distinction between laboratory or real-life setting). These recommendations appear directly relevant to our resolution of inconsistencies in the developmental research literature and a clear portrayal of the knowledge children do and do not bring to instruction.

The apparent tendency for children to avoid the distributional mode and instead interpret probabilistic situations in terms of singular events corresponds with Kahneman and Tversky's analysis of interpretative propensities in adults. These authors conjectured: "People generally prefer the singular mode, in which they take an 'inside view' of the causal system that most immediately produces the outcome, over an 'outside view,' which relates the case at hand to a sampling schema" (Kahneman & Tversky, 1982, p. 153). Kahneman and Tversky argued that although one can reason probabilistically in either form, the distributional form tends to support better assessments of probability.

Expected Distributions of Outcomes

It is difficult to locate convincing documentation of young children's grasp of expected distributions of outcomes in any but very simplistic forms and settings. Both Fischbein and Piaget appear to have identified rudimentary aspects of this construct in young children, in conjunction with particular situations in which this knowledge is reflected.

Fischbein framed his experimental procedure in terms of a series of marble shoots. The three simplest bifurcating marble shoots involved equiprobabilities, with marble shoots of the same length. The marble shoots he subsequently presented to his subjects were more complicated. After eliciting the children's ideas of what would happen when you dropped a single marble in the marble shoot, Fischbein asked questions focused on relative frequencies:

> If I drop a marble a lot of times, one after the other, will it come out at each place the same number of times, or will it come out at each place the same number of times, or will it come out of some places more often than others?

What makes you think that? [or] What did you think of, that made you say that? (Fischbein, 1975, p. 162)

Fischbein reported that all ages, from preschool up, correctly answered the top-level prediction question, in these simplest situations with equiprobabilities and marble channels of equal length. However, a failure to differentiate "objective conditions" from "subjective choices" was revealed in the justifications of the preschoolers, as reflected in statements such as the marble "goes where it wants." Fischbein also reported considerable hesitation in the school-age children interacting with marble channels of equiprobabilities. He attributed their discomfort to the uncertainty in the situation and their corresponding preference to somehow justify making a deterministic identification of which shoot the marble will go down.

It is when situations of unequal probability are introduced that the school-age children outperform the preschoolers and, according to Fischbein, become more comfortable with their answers. Thus in the marble shoot with one bifurcation of the marble path, followed by a second bifurcation in one of the two channels, the performance of the 7- and 8-year-olds surpasses the performance of the preschoolers.

Fischbein concluded that the performance of both the preschooler and 7- and 8-year-old cohorts breaks down in the situation with two levels of radially diverging pathways (with just 0% and 2% correct responses at the two age levels). This level of the task actually involves the mathematically demanding conjunction of probabilities, for example, with each path at the first level of divergence having a probability of $1/4$ and paths thereafter having probabilities from $1/3$ to $1/6$, thus resulting in conjunctive probabilities varying from $1/12$ to $1/24$.

The Piaget and Inhelder studies documented other forms of competence at the 7- to 9-year-old age span, although again under restricted task characteristics. Piaget and Inhelder claimed that, in reasoning about relative odds in two distributions, 7- to 9-year-old children simultaneously consider "the favorable cases (A), the unfavorable cases (A^1), and the totality of possible cases" (1975, p. 151). Furthermore, this age level can accurately assess the relative odds when either the number of favorable cases or the number of total cases is held constant across the collections under comparison. Thus:

SOU (8 years, 3 months) [Asked to designate which set he wanted to choose from, where one collection has one cross out of 3 and the other collection has 2 crosses out of 3] *Here is more likely because there are two crosses. That's more than one. I can get it twice here and only once there.* The experimenter asks, "And how about this [with 4/5 versus 3/5]." SOU responds, *Here, because there are 4 with crosses and only 3 there.* [In comparison of 1/5 with 1/4] *Here [1/4] it's more certain because there are only 4 counters. I'd lose fewer times than with five.* (Piaget & Inhelder, 1975, p. 150)

In Piaget's study, as well Acredolo's, the sizes of the sets are so small that the distributions or relative odds calculation is further simplified.

In this same study, Piaget and Inhelder concluded that 4- to 6-year-olds' successful assessment of relative odds is restricted to situations where perceptual analysis will suffice. For example:

> Gat [6 years, 2 months] expected to find a red counter mixed with two or three white ones more easily than the white ones, but when a fourth white one was added, he reversed his decision. At a certain point, the increasing number of white counters brings him to the decision that one of the white counters would be drawn more easily than the single red one. (Piaget & Inhelder, 1975, p. 144)

Piaget and Inhelder attributed whatever limited success these children have to "intuitive" or "perceptual" strategies, paralleling the Falk et al. (1980) conjecture of how their young subjects may have chosen the spinner or top with the most favorable odds.

The distinction between perceptual and conceptual basis for probability judgments appears fundamental. We can interpret some of the wide age discrepancies reported in the developmental research literature in terms of this distinction. Consider, for example, Davies' (1965) study, analyzing the differences in performance between what she referred to as nonverbal and verbal tests of probabilistic reasoning. Davies presented two tasks to children, 3 to 9 years of age. In the nonverbal procedure, she presented two gum-ball-like machines, side by side. One machine contained white and red marbles in the ratio of 4:1; the other machine contained white and red marbles in the reversed ratio.

The child's nonverbal understanding of probability was measured by the percentage of trials where the child chose the machine with the higher expected payoff of the targeted color. Fifty percent of the 4-year-olds and 69% of the 5-year-olds succeeded at this task. However, most of these children performed poorly on a task where they were asked both to indicate which color they thought they would get if they closed their eyes and drew from a particular box and to explain why they thought so. Indeed, Davies reported that no 4-year-old and just 31% of the 5-year-olds succeeded at this task. She noted that whereas the mean age for acquisition of the ability to explain their choice was 7 years, 4 months, the ability was not reflected by all subjects until 9 years. Although Davies acknowledged that her results are not sufficient to definitely identify what factor accounts for differential task difficulty, we focus on one of her hypotheses, "that the younger child is even unaware of the [probability] concept and of its influence on his behavior" (Davies, 1965, p. 787).

This tricky yet fundamental distinction between perceptual expectancies and probabilistic reasoning has also been identified in the adult judgment and decision-making literature. According to Kahneman and Tversky (1982), adults frequently rely on a perceptual level of sampling. These researchers argued that the key indicator of the more basic perceptual approach is the absence of any expression of doubt or uncertainty. Kahneman and Tversky asserted:

> The suppression of uncertainty and equivalence in perception suggest that we may be biologically programmed to act on the perceptual best bet, as if this bet involved no risk of error. A significant difference between the conscious experiences of perception and thought is that the latter can express doubt and uncertainty, while the former normally do not. (1982, p. 147)

A methodology that relies on best choices, such as has been used so frequently in the cognitive developmental research, cannot differentiate the perceptual from a more analytical approach, as it fails to consider whether or not the subject has manifested any expression of doubt or uncertainty.

In short, children's intuitions concerning relative distributions of possible outcomes seem much more limited than their intuitions of incertitude or the likelihood of an event. Although many studies in which the probabilities involve such comparisons have elicited apparently sensible choices from young children, close analysis of this research suggests that, instead of interpreting the tasks in terms of relative distributions, the children may well have avoided distributional thinking by either simplifying the task to one of assessment of relative likelihoods of specific outcomes or basing their choice on perceptual features of the collections.

It may well be that there are very few situations that arise in a child's everyday life that require an analytical distributive mode of reasoning. Indeed, this mode of thinking appears sufficiently complex that it may well demand some instructional support. A teaching experiment of Fischbein (1975) provides an optimistic note about the possibilities of children's understandings, at least at the upper end of the primary-grade age span.

Fischbein's instructional protocol emphasized the distributional nature underlying this form of probabilistic reasoning. For example, in one of the procedures presented to 9- to 13-year-olds, the child was presented with an urn containing two colors of beads. Fischbein reported that probabilities were determined for the different colors, presumably with considerable adult scaffolding. One color was designated as the child's winning color, the other as the losing color. The child iteratively drew and replaced a marble as the results were recorded. Results were chunked into series of 10 draws. The child's attention was directed to oscillations of the frequencies. The protocol incorporated key aspects of this form of probability, including: "(a) a theoretical prediction, (b) an experiment verifying the prediction, and (c) a demonstration of the oscillating frequencies, with their convergences toward a theoretical point, which is probability" (Fischbein, 1975, p. 92). The researcher reported that the children's interest was piqued by wanting to get the "winning color" and in wanting to find out whether or not their prediction was correct. Fischbein concluded that concrete operations (assumed on the part of the 9-year-olds) was sufficient for significant on-task learning of these ideas.

We need to interpret these findings within the perspective that education is fundamentally concerned with what children can accomplish with strategic instructional support. Most of these developmental studies had a minimal instructional component. Instruction, as Fischbein illustrated, can fundamentally change our view of what children can understand, including the core concepts of probabilities framed in the distributive mode.

ANALYSIS OF INSTRUCTIONAL CHALLENGES

Conceptual Challenges

We can address the problem of what conceptual lacunae continue to pose problems for primary-grade children, by revisiting David Moore's definition of randomness, the core construct underlying chance and probability. Again: "Phenomena having uncertain individual outcomes but a regular pattern of outcomes in many repetitions are called *random.* 'Random' is not a synonym for 'haphazard,' but a description of a kind of order different from the deterministic one" (Moore, 1990, p. 98). While primary-grade children's problem solving reflects important aspects of this construct, key aspects are also weak or missing.

As revealed in the preceding analyses, primary-grade children have some understanding of the incertitude and unpredictability of events. They combine this idea with predictions of outcome, in the specific sense of the likelihood of a given event. The conceptual integration of unpredictability and patterns in the form of distributions appears more limited. Second, the unschooled primary-grade children's conceptions fail to reflect the mathematician's framing of both randomness and probability over large numbers of events. Some sense of infinity, or at least the Law of Large Numbers, is fundamental to a robust conceptualization of randomness, particularly in how it applies to frequentist probabilities. Thus instruction will presumably need to address the subtle idea of phenomena whose individual outcomes may be undetermined but that reveal patterns of outcomes over the long haul.

Building From Children's Solid Intuitions

This chapter has focused on the identification of intuitions, from the perspective that instruction grounded in children's intuitions has a greater potential of sensibility and purpose. In the words of the NCTM document *Setting a Research Agenda:*

> The learning of mathematics is not just the acquisition of knowledge in the subject matter, nor even just the construction of that knowledge using general methods of learning and understanding. Instead, it is the process of building upon the concepts and principles that children already possess when they encounter the subject matter of mathematics, and then *organizing and restructuring those concepts and principles.* (NCTM, 1989b, p. 33; emphasis added)

Although we may begin with the intuitions, instruction clearly needs to transcend them.

As the NCTM emphasized, one crucial dimension of this process involves the connecting and structuring of intuitions, such as the integration of ideas of incertitude and patterns of expected outcomes. In the process come conceptual elaborations, such as the extension from framing expectations in terms of the likelihood of given events to distributions as well, and the conceptualizations of patterns in a small data set to the idea of patterns over the long haul.

A second dimension in this process of building from intuitions concerns the meta-level of the knowledge. Apart from the instructional process, intuitions are typically unexamined. Pylyshn (1978) defined *reflective access* as "one's ability to mention as well as use." In a somewhat more stringent conceptualization of the same basic idea, Campione (1984) defined the same term as where "the individual possesses not only fact X, but also the notion of the fact X" (p. 120). The scaffolding of reflective access, and of the critical analysis and conceptual elaboration that can build on such explicitation, constitutes a key challenge in the success of young children's statistical instruction.

To "Have" or to "Use": The Challenge of Appropriate Application

Although the developmental literature has been largely framed in terms of do they or don't they "have" the construct yet, other literatures point to the fundamental challenge of appropriate application. Analysis of intellectual history (Gigerenzer, Switjink, Daston, Beatty, & Kruger, 1989; Hacking, 1990) reveals this issue has long challenged the development of the natural and social sciences. Examination of the judgment and decision-making literature documents the considerable challenge it poses to contemporary adults, including those with statistical training (e.g., Kahneman, 1991; Kahneman et al., 1982). Both these literatures document a general tendency to err on the side of attributing too much to deterministic causality.

Our successful development of children's statistical and probabilistic intuitions demands not only the scaffolding of key conceptual constructs, but also their appropriate application. Alongside conceptual elaboration, appropriate utilization needs to constitute a primary focus in teachers' scaffolding of children's interpretation of variability in data. Success in this realm will involve both conceptual constructions and epistemological changes, for the propensity to consider chance is actually a form of epistemological orientation. In short, if children believe some form of determinism underlies events, they will be little inclined to consider the possibility that the variability they observe is due to chance. Application of chance assumes a view of the world with a place for chance events.

SUMMARY

The mathematics reform movement has called for incorporation of statistics and probability, beginning at the primary-grade level. More specifically, national curricular documents of the United States, the United Kingdom, and Australia have recommended that these young children explore concepts of chance and probability and conduct empirical investigations, including the posing of their own questions and the collection, organization, and interpretation of data. This chapter analyzed the research literature to identify children's intuitions on which this curricular agenda could build, as well challenges it will confront.

The chapter identified five key intuitions that primary-grade teachers could build on in teaching statistics and probability to young children. These include relative magnitude, part–whole relations, incertitude and indeterminacy, the likelihood of an event, and, in much more limited form, expected distributions of events. Three key challenges of statistics and probability instruction at this age level were also identified. First, from a conceptual perspective, this age group evidenced limited integration of ideas of incertitude with patterns, especially the idea of patterns emerging over the long haul. Second, although accessible instruction needs to be grounded in children's intuitive understanding, it needs to transcend them. Instruction needs to scaffold the explicitation of these intuitions, as it helps the children refine, elaborate, and restructure these ideas. Finally, instruction in this sphere needs to focus not only on the considerable challenge of conceptual elaboration, but also on issues concerning when and where the chance frame is appropriate to use.

ACKNOWLEDGMENT

The author thanks Jan LaTurno for her constructive comments on an earlier version of the chapter.

REFERENCES

Acredolo, C., O'Connor, J., Banks, L., & Horobin, K. (1989). Children's ability to make probability estimates: Skills revealed through application of Anderson's functional measurement methodology. *Child Development, 60,* 933–945.

American Association for the Advancement of Science. (1993). *Benchmarks for scientific literacy.* New York: Oxford University Press.

American Statistical Association–National Council of Teachers of Mathematics Joint Committee on the Curriculum in Statistics and Probability. (1985). *Teaching of statistics within the K–12 mathematics curriculum.* Washington, DC: American Statistical Association.

Anderson, H. H. (1980). Information integration theory in developmental psychology. In F. Wilkening, J. Becker, & T. Trabasso (Eds.), *Information integration by children* (pp. 1–45). Hillsdale, NJ: Lawrence Erlbaum Associates.

Bassano, D. (1985). Modalités de l'opinion: Quelques expressions de la croyance et de la certitude et leur differentiation entre 6 et 11 ans. *Cahiers de Psychologie Cognitive, 5*(1), 65–87.

Bullock, M. (1985). Causal reasoning and developmental changes over the preschool years. *Human Development, 28,* 169–191.

Burbules, N. C., & Linn, M. C. (1991). Science education and philosophy of science: Congruence or contradiction? *International Journal of Science Education, 13*(3), 227–241.

California State Department of Education. (1985). *California State Framework for Mathematics.* Sacramento, CA: Author.

Campione, J. C. (1984). Metacognitive components of instructional research with problem learners. In F. E. Weinert & R. H. Kluwe (Eds.), *Metacognition, motivation, and learning* (pp. 109–132). West Germany: Kuhlhammer.

Carpenter, T. P., Ansell, E., Franke, M. L., Fennema, E., & Weisbeck, L. (1993). Models of problem solving: A study of kindergarten children's problem-solving processes. *Journal for Research in Mathematics Education, 24*(5), 428–441.

Cobb, P., Yackel, E., & Wood, T. (1993). Theoretical orientation. In Rethinking elementary school mathematics: Insights and issues. *Journal for Research in Mathematics Education,* Monograph No. 6, 21–32.

Davies, C. M. (1965). Development of the probability concept in children. *Child Development, 36,* 779–788.

Dossey, J. A., Mullis, I. V. S., Linquist, M. M., & Chambers, D. L. (1988). *The mathematics report card: Are we measuring up. Trends and achievement based on the 1986 National Assessment.* Princeton, NJ: Educational Testing Service.

Falk, R., Falk, R., & Levin, R. (1980). A potential for learning probability in young children. *Educational Studies in Mathematics, 11,* 181–204.

Fischbein, E. (1975). *The intuitive sources of probabilistic thinking in children.* Dordrecht, The Netherlands: Reidel.

Gelman, R., & Gallistel, C. R. (1978). *The child's understanding of number.* Cambridge, MA: Harvard University Press.

Gigerenzer, G., Swijtink, Z., Daston, L., Beatty, J., & Kruger, L. (1989). *The empire of chance: How probability changed science and everyday life.* Cambridge, England: Cambridge University Press.

Green, D. (1988, August). Children's understanding of randomness. In R. Davidson & J. Swift (Eds.), *Proceedings of the Second International Conference on Teaching Statistics.* Victoria, British Columbia. Victoria, BC: University of Victoria.

Hacking, I. (1990). *The taming of chance.* Cambridge, England: Cambridge University Press.

Hawkins, A. S., & Kapadia, R. (1984). Children's conceptions of probability: A psychological and pedagogical review. *Educational Studies in Mathematics, 15,* 349–377.

Howell, W. C., & Burnett, S. A. (1978). Uncertainty measurement: A cognitive taxonomy. *Organizational Behavior and Human Performance, 22,* 45–68.

Huber, B. L., & Huber, O. (1987). Development of the concept of comparative subjective probabilities. *Journal of Experimental Child Psychology, 44,* 304–316.

Kahneman, D. (1991). Judgment and decision making: A personal view. *Psychological Science, 2(3),* 142–145.

Kahneman, D., Slovic, P., & Tversky, A. (1982). *Judgement under uncertainty: Heuristics and biases.* Cambridge, England: Cambridge University Press.

Kahneman, D., & Tversky, A. (1982). Variants of uncertainty. *Cognition, 11,* 143–157.

Klahr, D., & Wallace, J. G. (1976). *Cognitive development: An information processing view.* Hillsdale, NJ: Lawrence Erlbaum Associates.

Konold, C. (1991). Informal conceptions of probability. *Cognition and Instruction, 6(1),* 59–98.

Kuzmak, S., & Gelman, R. (1986). Young children's understanding of random phenomena. *Child Development, 57,* 559–566.

Metz, K. E. (1995). Re-assessment of developmental assumptions in children's science instruction. *Review of Educational Research, 65(2),* 93–127.

Metz, K. E. (in preparation). *Children's interpretation of random phenomena.*

Moore, C., Pure, K., & Farrow, D. (1990). Children's understanding of the modal expression of speaker certainty and uncertainty and its relation to the development of a representational theory of mind. *Child Development, 61,* 722–730.

Moore, D. S. (1990). Uncertainty. In L. A. Steen (Ed.), *On the shoulders of giants: New approaches to numeracy* (pp. 95–137). Washington, DC: National Academy Press.

National Council of Teachers of Mathematics. (1989a). *Curriculum and evaluation standards for school mathematics.* Reston, VA: Author.

National Council of Teachers of Mathematics. (1989b). *Setting a research agenda.* Reston, VA: Author.

Piaget, J., & Inhelder, B. (1975). *The origin of the idea of chance in children.* London: Routledge & Kegan Paul.

Piaget, J., Inhelder, B., & Szeminska, A. (1965). *The child's conception of number.* New York: Basic Books.

Pylyshn, Z. W. (1978). Computational models and empirical constraints. *Behavioral and Brain Sciences, 1,* 93–100.

Resnick, L. B. (1983). A developmental theory of number understanding. In H. P. Ginsburg (Ed.), *The development of mathematical thinking* (pp. 109–151). New York: Academic Press.

Resnick, L. B. (1986). The development of mathematical intuition. In M. Permutter (Ed.), *Perspectives on intellectual development: The Minnesota Symposium on Child Development, 19* (pp. 159–194). Hillsdale, NJ: Lawrence Erlbaum Associates.

Resnick, L., Lesgold, S., & Bill, V. (1991). From protoquantities to number sense. *Proceedings of the North American Psychology of Mathematics Education–North America,* pp. 305–311.

Riley, M., Greeno, J., & Heller, J. (1983). Development of children's problem-solving ability in arithmetic. In H. P. Ginsburg (Ed.), *The development of mathematical thinking.* New York: Academic Press.

Romberg, T., Allison, J., Clarke, B., Clarke, D., & Spence, M. (1991, November). *School Mathematics Expectations: A comparison of curricular documents of eight countries with the NCTM Standards of the U.S.* Paper prepared for the New Standards Project.

Rosier, M. J., & Keeves, J. P. (1991). *The I.E.A. Study of Science, 1: Science education and curricula in twenty-three countries.* Oxford, England: Pergamon.

Saxe, G. (1991). *Culture and cognitive development: Studies in mathematical understanding.* Hillsdale, NJ: Lawrence Erlbaum Associates.

Siegler, R. S. (1981). Developmental sequences within and between concepts. *Monographs of the Society for Research in Child Development, 46(2,* Serial No. 189), 1–74.

Siegler, R. S., & Robinson, M. (1982). The development of numerical understandings. In H. W. Reese & L. P. Lipsitt (Eds.), *Advances in child development and behavior* (pp. 241–312). New York: Academic Press.

Steen, L. A. (1988). The science of patterns. *Science, 240,* 611–614.

Steinbring, H. (1991). The theoretical nature of probability in the classroom. In R. Kapaidia & M. Borovcnik (Eds.), *Chance encounters: Probability in education* (pp. 135–167). Dordrecht, The Netherlands: Kluwer.

Vallone, R., & Tversky, A. (1985). The hot hand in basketball: On the misperception of random sequences. *Psychological Review, 90,* 293–315.

Yost, P. A., Siegel, A. E., & Andrews, J. M. (1962). Nonverbal probability judgments by young children. *Child Development, 33,* 769–780.

CHAPTER SEVEN

Developing Middle-School Students' Statistical Reasoning Abilities Through Simulation Gaming

Sharon J. Derry
Joel R. Levin
University of Wisconsin–Madison

Helena P. Osana
University of Missouri

Melanie S. Jones
University of Wisconsin–Madison

We are exploring the use of complex simulation gaming as a pedagogical strategy for developing students' statistical reasoning capabilities in the context of realistic problem-solving situations. Our work is motivated by several different strands of research and theory. First, we are attempting to address problems revealed by recent studies showing that many professional and lay adults in mainstream American society are not proficient at reasoning probabilistically or statistically about important societal issues. This is the case even though literacy and informed decision making today require at least informal (if not formal/computational) probabilistic and statistical thinking (e.g., Konold, 1989; Kuhn, 1991; Tversky & Kahneman, 1974). Second, in response to recent widespread interest in situated cognition (e.g., Lave, 1991), we are testing an instructional approach that reflects the theoretical perspective of Vygotsky (e.g., 1978). Vygotsky argued that transmission of scientific thought to broader cultures requires particular forms of social interaction, recently called "cognitive apprenticeship" (e.g., Collins, Brown, & Newman, 1989). Our instructional philosophy also borrows from: (a) the teachings of Dewey (1938), who advocated meaningful school practice that extends adult society, and (b) radical constructivism (e.g., von Glasersfeld, 1990), which values educational practice based on reflective activity—rather than idea transmission—in which students build up new knowledge and understandings through social and physical interaction with the instructional environment.

The educational programs we are designing and investigating engage students in the following types of mentored activities: (a) reflective discussions of popular media presentations (e.g., movies, television shows, newscasts) that illustrate and require statistical reasoning, and (b) extensive "hands-on" participation in simulations of authentic community and professional activities (e.g., conducting research, participating in hearings and conferences, etc.) that require development, presentation, and critique of statistical arguments about important societal issues. Both media-anchored discussions and authentic simulations are designed to serve as realistic rich contexts for mentoring students' ways of thinking. The goal for students during these activities is to develop and improve their use of evidence and statistical reasoning in argumentation. For example, students in our college course on statistical reasoning enhance their understandings of experimental design, statistical inference, and scientific research in general as they conduct experiments on fast-growing plants and present their findings at a "conference." They develop new understandings of bias in employment and admissions practices through a simulation game in which they investigate employment practices of fictional school districts, using statistical concepts such as the sampling distribution of a proportion. This type of college course is unusual because the major mode of classroom instruction is mentored activity rather than the standard lecture (see Derry, Levin, & Schauble, 1995).

This chapter focuses on our work in middle schools. We describe the implementation and evaluation of a 3-week instructional unit that was adapted to three different subject-matter contexts (science, social studies, and mathematics). In all contexts, the purpose of the unit was to improve students' abilities to think and reason statistically about real-world issues. Prior to instruction, teachers and researchers collaborated together in planning and in-service training sessions. During the 3 weeks of instruction, students watched and discussed (with researchers and teachers as guides) a popular movie, *Lorenzo's Oil*, an emotional drama that focuses on such issues as ethics in medical research, experimental control, informed single-case observations versus randomized clinical trials, and governmental regulation of the scientific and lay community. Following the movie, students then carried out a lengthy mock legislative hearing dealing with government regulation of the dietary supplement and vitamin industry, an issue that was present in the news of the time and that shared several parallels with the story line in *Lorenzo's Oil*. During the 3 weeks of instruction, teachers and researchers made several short presentations on the topics of thinking, presenting arguments, and statistical decision making. However, there were no homework assignments pertaining to statistical reasoning and a relatively small amount of time (1 day) was devoted to direct instruction on that topic. Nevertheless, it was intended and expected that students' statistical reasoning would measurably improve as a result of participation in mentored discussions and activities.

To evaluate student performance gains in their ability to reason statistically in everyday (and emotional) contexts, two assessment tasks and accompanying scoring rubrics were devised. The assessment tasks presented students with dialogue from fictional dramatic trials, inspired by popular television shows, in which witnesses and lawyers presented evidence, arguments, and counterarguments to support their cases. For each trial scenario, the students' task was to analyze the dialogue, draw a conclusion, and support the conclusion with reasons. Students' responses were judged by scorers who were trained to use a rubric and framework that were developed by project researchers and based on current research on statistical reasoning. Taken together, the rubric and framework amount to a normative model for good thinking that both prescribes what is valued in student reasoning and proscribes what is not. Because the tasks and scoring procedures were not designed as project-specific instruments, but rather as general assessment tools to be shared with teachers and researchers who want to judge students' reasoning abilities within various instructional contexts, they are described in some detail within this chapter.

A CLASSROOM-BASED STUDY OF STATISTICAL REASONING INSTRUCTION

Invitations to participate in a 3-week instructional unit were issued to teachers of "home-base" classes within a racially and socioeconomically diverse middle school in the Midwest. Home-base classes combine a regular 50-min subject-matter period with a 20-min homeroom period, creating a 70-min class period. Home-base classes were chosen because they were not formed by ability tracking and also tended to be equally heterogeneous with respect to students' race and socioeconomic status. Two teachers, a science teacher and a social studies teacher, agreed to participate in the instruction. (One seventh-grade mathematics class also participated but was not included in the study proper.) In addition, eight eighth-grade home-base classes that did not participate in the instruction agreed to take pretests and posttests, thus enabling us to compare nonparticipating and participating classes' gains in statistical reasoning ability. Both instructional and comparison classes were given pretest and posttest measures based on the performance assessment tasks previously mentioned, which are described in a later section. Video data also were collected in the two instructional classes, to permit qualitative analyses of classroom interactions. However, results reported here are based exclusively on the performance data.

Thinking About Science and Government Regulation

The instructional unit was implemented in the spring of 1994. The activity spanned a 3-week period, during which two eighth-grade teachers each offered one class period every day. Class periods lasted 70 min and incorporated both homeroom class (normally 20 min) and the subject-matter class (either social studies or science, 50 min) that is normally taught just before homeroom class.

During the first 4 days of the instructional unit, the students watched the film *Lorenzo's Oil* in their classrooms. The movie depicts a true story of the Odone family, who learn that their 5-year-old son, Lorenzo, has a rare terminal disease (adrenoleukodystrophy, or ALD). Despite the prognosis, the Odones set out to save their child by discovering a cure, colliding with doctors, scientists, and support groups who are reluctant to encourage the couple in their quest. The film depicts numerous conflicts and raises many questions concerning governmental regulation and the differences between expert and lay statistical and scientific reasoning. This film served as a context for introducing various topics related to statistical influence, medical research, and scientific reasoning. Laser disc technology was used, which allowed for easy access to critical scenes and segments during discussions.

Students were placed in small groups to discuss important issues raised in the film. At the end of each day's viewing, the groups were provided with one-page worksheets that focused on an important question or issue related to the movie. These questions and worksheets were designed to stimulate statistical reasoning in group discussion. Students also were asked to share their responses and reactions with others during a general class discussion. During these discussions, a Socratic teaching method was used by the teachers and researchers. One of the group discussion guides is given as an example in Appendix A.

On day 4, an outside expert (a scientist who had consulted on the movie) spoke in each class and answered students' questions concerning the research and social issues depicted in the film, as well as the researcher's personal experiences in making the film.

On days 5 through 7, rotating 70-min, whole-class instructional presentations were given in each class on the following topics: making presentations more effective with models, graphs, and charts; credible scientific research and valid statistical inference; and thinking as evidential argument. Direct instruction was used primarily, although class discussions and activities were also were employed.

The unit on presentations was taught by the eighth-grade science teacher in collaboration with a project researcher. They discussed ways in which models, graphs, charts and illustrations can be used as tools to illustrate an idea or a piece of evidence. With the assistance of the microcomputer package, Claris Works, an introduction to simple graphing and drawing packages on the computer was given. The second instructional unit, taught by a project researcher, was a demonstration–presentation on scientific research, errors associated with statistical inference, research methodology, and probabilistic reasoning. The third unit, taught by the eighth-grade social studies teacher in collaboration with a project researcher, used lecture, a humorous video (made in-house), and whole-class discussion to illustrate the differences between convincing and unconvincing evidential argument.

A simulation game began on day 8 of the intervention and lasted until the end of the 3 weeks. The goal of the game for the entire class was to prepare and conduct a legislative hearing. The purpose of the hearing was to decide whether a legislative bill that proposed governmental controls on advertising and availability of vitamins and dietary supplements should be accepted, rejected, or modified. During the game, students worked in their small groups to prepare arguments. Each group was given a large envelope in which were placed role descriptions for group members, an overview text describing the general status of the vitamin regulation debate, and various journal articles and government reports (all modified for readability) on research findings pertaining to benefits and dangers of dietary supplements. Within each group, students assumed fictional roles representing various members of the affected real-world community. Some students represented those with vested interest in the issue of vitamin regulation: store owners, members of the Food and Drug Administration, vitamin consumers, members of the health profession, medical researchers, and vitamin manufacturers. Other students were assigned roles representing more balanced positions on the issue: journalists, television and radio reporters, state legislators, and so forth. Planned role distributions resulted in learning groups that, as a whole, represented special interests and were expected to be either supportive of major government regulation, reluctant to regulate, or relatively neutral with respect to regulation. In this way the stage was set for groups and group arguments to represent opposing sides at the hearing.

On days 8 through 11, using the materials provided, the students analyzed data, developed arguments, and designed presentations to support their positions on the regulation issue. Any rules for collaborative work were established by the students themselves within their groups. However, the role of the project researchers and teachers was important at this stage of the intervention, and each had previously participated in workshops on mentoring and statistical reasoning in preparation for these roles. In each of the two experimental classrooms, researchers and teachers "floated" from group to group, mentoring student activity. The mentors modeled good reasoning, explicitly explained statistical ideas when necessary in the context of group issues, and generally helped students stay on track and work collaboratively. They constantly encouraged students to employ evidence and statistical reasoning in their arguments. Teachers also periodically led organizing class discussions that required students to summarize and analyze their progress each day.

On days 12 through 15, the students presented and defended their pre-pared arguments during a 4-day mock legislative hearing. In each class, the rules for conducting the hearing were proposed by the group that served as the legislative committee. Rules determined how long speakers would be allowed to speak, who would be allowed to question speakers, and when and how much rebuttal would be allowed. The legislative committee also conducted and monitored the hearing, and was responsible for making sure hearing rules were followed. After each group presented their arguments either in favor of or in opposition to regulation, the students in legislative roles deliberated and presented to the class a reasoned decision regarding the fate of the bill.

After the intervention was completed, approximately 20 to 30 min were spent in each class discussing with the students their experiences, reactions, and attitudes toward the simulation. Students' evaluations of the 3-week activity were collected using anonymous questionnaires.

Assessment Tasks

The impact of the instructional unit on students' reasoning abilities was evaluated using two assessment tasks. Each of the two tasks consisted of a three-page printed dialogue, followed by questions related to the dialogue. The dialogues were scenarios (adapted from a popular television program) in which various personalities argued in a court of law about a social issue. Both sides in the debate presented various forms of research evidence and counterarguments.

For example, one form of the test, called John's Trial, described a scene in which a young boy is being tried in adult court for violently harming a friend. The lawyer defending him claims the boy is not responsible for his actions because he possesses an extra Y chromosome. She presents evidence suggesting that the extra Y chromosome is related to violent behavior, evi-dence that is based on studies conducted in the 1960s and 1970s, which the prosecuting attorney argues were inconclusive. The other form of the test, the Andrevil Trial, told the story of Kenny, a young boy with a severe form of cancer. He is treated with a drug called Andrevil as part of an experimental trial. In the court scene, the boy and his mother are suing to stop the research involving other children being carried out by the doctor who administered the drug to Kenny, because Kenny's condition had worsened over the course of the experiment.

Each assessment dialogue was followed by two questions that required students to reason about the material in the dialogue. Providing an appro-priate answer to one of the questions required the application of statistical reasoning. However, statistical reasoning was neither required nor appro-priate for the second question. The purpose of having both a statistical and a nonstatistical question was to ascertain whether participation in the instructional unit resulted in a tendency for students to try to apply statistical ideas even when they are not appropriate (i.e., to overgeneralize those ideas). The ability of students to discriminate between statistical and nonstatistical reasoning situations was considered important.

For example, the statistical question following the Andrevil Trial scene asked whether or not Dr. Birch should be allowed to continue his research and why, based on evidence in the dialogue. The nonstatistical question asked whether academic researchers should in general be permitted to accept money from drug companies. Students were asked to answer these questions and then to provide up to three statements for each question that would justify their answers. A copy of the Andrevil Trial version of the test is provided as Appendix B.

Assessment Administration

There were two forms of the assessment task: Andrevil Trial and John's Trial. Before the instructional unit, half the students (selected randomly) received one form and the remaining students received the other. Approximately 10 days after the intervention was completed in the two experimental classes, postassessments were administered. The same instruments administered during the pretest phase were used (John's Trial and Andrevil Trial), with those students who received John's Trial on the pretest receiving the Andrevil Trial on the posttest, and vice versa.

Collaborative Versus Individual Testing. Because the instructional intervention was in part intended to help build cooperative reasoning skills, we tested collaborative student performance. Students within classes were assigned by teachers to small testing and instructional groups of four to five students each. In the comparison classes, the testing groups were formed through random assignment.

To permit us to compare collaborative with individual reasoning performance, one group from each class was selected at random and sent to the library during testing times. Students in the library were tested individually. They were asked to think about their story dialogue and to answer their questions individually without consulting peers. By contrast, students in the small-group condition were tested in their classrooms and discussed the test and their answers with their peers before they wrote them individually.

Individual Testing. Individually tested students were given 12 min to read the court scenario and think about the two questions given at the end of the scene. They were told to formulate their responses and were allowed to make notes. The participants were subsequently given a total of 5 min to write their three "why" responses for each question, and were paced to allow adequate time for both questions. They were reminded to work in silence, and to refrain from communicating with each other.

Group Testing. After one group from each class was sent to the library, four groups remained in each of the 10 classes, two receiving the Andrevil Trial version of the test and two receiving John's Trial. It was intended that all groups receiving the Andrevil Trial story at pretest would receive the John's Trial story at posttest, and vice versa. (An administrative error resulted in some groups receiving the same version at both pretest and posttest times, and these groups were excluded from the statistical analyses.)

Students in the group testing were told to read the story as a group and to discuss the issues raised by the questions at the end. They were given 12 min to read and discuss the story and questions, and 5 min to respond individually to the questions. Thus, the time limits that were imposed matched those imposed on the individually tested group.

Scoring of Assessment Tasks

Students' responses to the statistical and nonstatistical "why?" questions following the Andrevil Trial and John's Trial stories were scored according to a rubric based on response categories devised by the researchers. These categories, shown in Table 7.1, represent types of desirable and undesirable thinking that might be found in student answers. Table 7.1 is based on a normative model for good thinking derived from recent research on statistical reasoning, decision making, and argumentation (e.g., Konold, 1989; Kuhn, 1989; Nisbett, Fong, Lehman, & Cheng, 1987; Shaughnessy, 1992; Tversky & Kahneman, 1974). It is intended as a general grading scheme that will be used in our future teaching and research.

As examples, categories were assigned for using correlational reasoning (A), for considering degree of strength of relationships (B), and for citing specific counterpositive evidence (C). Categories were also developed for recognizing the desirability of large sample sizes in relation to the issue of seeing chance versus real relationships (D), for recognizing that correlation need not imply causation (E), for taking into account methodological quality of the research (F), for noting that further research (replication) is required before conclusions can be drawn (G), and for raising risk/benefit issues (M). Categories were also developed to reflect inappropriate deterministic thinking (I) and unsubstantiated opinion (L), among others—see Table 7.1 for the complete list of categories, along with descriptions and examples.

The awarding and subtraction of points based on these categories should vary in different situations, depending on the goals of instruction and the story used in the reasoning task. For some story scenarios, certain ideas might be

TABLE 7.1

Scoring Categories

John's trial: independent variable = Y chromosome
                  dependent variable = violent behavior
Andrevil trial: independent variable = Andrevil
                  dependent variable = condition of cancer patients

A. *Relationship between two variables*

Does protocol either explicitly or implicitly use in its reasoning the idea that the independent variable (Y chromosome, Andrevil) may or may not be related to the dependent variable (violent behavior, condition of patients)?

Examples:

J: It may be true that the extra Y chromosome can lead to violent behavior.

A: Some kids are getting better. There is a chance the drug could work, and save the rest of the kids' lives.

B. *Strength of relationship*

Does protocol use in its reasoning that the relationship between the independent variable (Y chromosome, Andrevil) and the dependent variable (violent behavior, condition of patients) can vary in strength?

Examples:

J: Some kids with the extra Y chromosome are violent, not all. We need to know how often the Y chromosome means violence.

A: The research must continue because the drug might not work on some, but it might work on others. The question is, "How often?"

C. *Counterpositive reasoning*

Does protocol indicate that there might be specific instances that *weaken* the relationship between the independent and the dependent variable?

Examples:

J: It is possible that not all people with the extra Y chromosome will act violently.

A: Almost half of the kids are dying or getting worse. This could mean that the drug doesn't work very well.

D. *Real versus chance relationship*

Does protocol indicate that the relationship between the independent variable and the dependent variable might be due to chance? Does protocol indicate that controlled experiments conducted repeatedly are needed to determine (with reliability) whether or not a real relationship exists between the two variables? Does response indicate that a single study might point to a chance (as opposed to a real) relationship?

Examples:

J: The relationship that was found between the extra Y chromosome and violence was a fluke.

A: One small study can't tell us whether the drug is good or not.

E. *Correlation does not imply causation/ Possibility of a spurious correlation*

Does the response reveal an understanding that even though the extra Y chromosome is linked to violent behavior, the Y chromosome is not necessarily the *cause* of violent behavior (or that Andrevil does not necessarily *cause* an improvement in, or worsening of, patient condition)? Does the response indicate that the correlation might in fact be due to some other confounding variable(s)?

Examples:

J: Kids who are violent may be that way because of the way they are raised, not because of an extra Y chromosome.

A: Maybe the Andrevil is not causing the kids to feel better; maybe the kids who are feeling better are on a special diet or something.

(Continued)

TABLE 7.1
(*Continued*)

F. *Quality of the research*

Does the protocol criticize any aspect of the research presented in the script (the research that tried to establish a link between Y chromosome and violence; the research that tried to establish a relationship between Andrevil and the condition of cancer patients)? Does the response indicate the necessity of doing proper research?

Examples:

J: There is not good evidence because the study about the extra Y chromosome is not finished yet.

A: There is no control group in the study and also the group is too small for accurate conclusions.

G. *Research evidence versus opinion*

Does the protocol indicate that research is valued as a way for answering important questions? Does the response suggest that further research is needed before important decisions can be made?

Examples:

J: This information needs to be researched more before you can say that there is a link between the extra Y chromosome and violence.

A: If you stop the research, then you can't find out whether Andrevil works or not.

H. *Citing expert testimony*

Does the statement indicate reliance on the importance of an expert's knowledge or testimony (without any justification)?

Examples:

J: They can control themselves, as Dr. 1 said.

A: I think the research should continue because Dr. Birch thinks more studies are needed.

I. *Deterministic reasoning*

Does protocol indicate that there might be some way (or will be some way in the future) to determine with certainty that the extra Y chromosome makes people violent, or that Andrevil cures cancer?

Examples:

J: It hasn't been proved yet that extra Y chromosomes causes them to be more violent.

A: If he does more research, he'll be able to prove that Andrevil works. Then cancer patients will be cured.

J. *Uncertainty of the effectiveness of the counseling treatment (John's trial only)*

Does protocol indicate recognition that there is uncertainty about the worth of the suggested treatment for boys?

Example:

J: The special school might not work. I think they should go to counseling instead.

K. *Wrong evidence cited*

Does protocol indicate a misinterpretation or an incorrect understanding of the evidence presented in the script (even though correct reasoning may be present)?

Examples:

J: There have not been any studies that show a link between the extra Y chromosome and violence.

A: The research must continue because over three-quarters of the patients are feeling better.

L. *Nonevidence*

Does the protocol justify conclusion with a) personal experience, b) nonscientific background knowledge, c) emotion, or d) unsubstantiated opinion? Does the protocol use in its reasoning ideas that are a) not logical, or b) incomprehensible?

(*Continued*)

184

TABLE 7.1
(*Continued*)

Examples:

J: If the kids were placed away from a lot of their peers, then when they returned it might shock them into committing another crime.

A: The research must stop because Dr. Birch is a serial killer.

M. Risk/benefit trade-off

Does the protocol indicate that the suggested treatment for boys (John's trial) is in some way too risky (too expensive, too dangerous) for what it is worth? Does the protocol indicate that it is too risky to continue research on Andrevil (because it may harm more patients than help, or because it is too costly)?

Or on the contrary, does the protocol indicate that the special treatment for boys (or continued Andrevil research) may do more good than harm?

Examples:

J: There would then have to be more schools for all kinds of special kids, and this would be too expensive.

A: I know it sounds cold but, the good of the many outweigh the needs of the few.

Z. No response

---

very important for a valid interpretation, whereas others might be less important. For example, consider the case in which instruction focuses on the concept of risks versus benefits and the teacher selects an assessment story for which a valid interpretation requires the use of risk/benefit concepts but not correlational concepts. In this case, the presence of risk/benefit thinking in a student's response should be rewarded with points, whereas absence of such thinking should be penalized. Moreover, use of correlational concepts should be penalized if their use is not appropriate for the story.

For this study, we developed a scoring system that awarded points for categories deemed equally appropriate for both test versions. The selection and weighting of categories were decided prior to analyzing data, with point values based on the researchers' consensus view of the category's importance to the scenarios. The appropriateness of the chosen scheme for the two tests is illustrated using two composite "ideal" student arguments, two for the John's Trial story and two for the Andrevil Trial story. These composite arguments tie together comments actually made by a number of different students. Each idealized argument is provided later in this chapter, followed by the categories that were used in scoring that argument and an explanatory rationale for how each category was applied in describing the given argument. Points assigned to the various categories were as follows: 2 points for categories B, C, D, E, and F; 1 point for categories A, G, and M; 0 points for categories H, I, J, and L; and minus 1 point for category K.

We note that the composite ideal arguments that follow are provided to illustrate the coding rationale only and do not provide an adequate picture of actual student performance on our test. Students in our study were limited to providing up to three statements of rationale for each question, and were not allowed time to develop full discursive explanations. This testing procedure was followed to minimize confounding effects of verbal ability and to conserve testing time. The scoring procedure we used assumes that actual students' responses represent a sample of the reasons they might cite in a more complete explanation.

Response Classifications Explained, Along With Composite Student Arguments

Andrevil Trial. Assume that the student agrees that Dr. Birch's research must stop. A possible line of reasoning might be as follows:

> It is too great a risk to continue the research because the percentage of children that are dying and getting sicker is too great compared to the group that is improving. Thus, there is too much risk associated with further research. Also, there is no randomized control group in the study and the experimental group to date is too small for conclusions to be reached. The children taking Andrevil could be getting better for other reasons. Thus, Dr. Birch should continue his research on animals, where better controls can be satisfied so that human lives are not at stake.

Based on the descriptions and letter codes given in the previous section, the classification rationale for the argument just given can be explained as follows:

The crux of this argument is that although there may be a positive relationship between Andrevil and getting better (A), the counterpositive evidence (C) is too strong to be ignored so that the risk to children appears too great (M). The children taking Andrevil in Dr. Birch's study may be getting better spontaneously or for reasons other than the Andrevil (E). Dr. Birch's research program is not of sufficient quality to warrant continued support at this time (F). Further research is needed (G) as we can determine the strength of the real relationship (B) only with larger studies over the long run (D), with better designed studies (F) with lower risk factors (M), such as with animal research.

Points for each category represent the researchers' collective opinion of which argument components should be given positive or negative values, and which are most or least important. In the present analysis, points were assigned only once per question. For example, even if two justification statements for the statistical question both cited counterpositive evidence (C), only a single 2-point C response was counted. After coding a student's response, we computed the student's weighted composite score for each question based on the sum of the points associated with each category given the response.

A similar example is now given for an argument justifying the opposite conclusion: that Dr. Birch's research must continue. The following justification might be offered:

> Even if people are dying or getting sicker while taking the drug, there are still kids that are getting better. In the long run, many more lives might be saved, so we need to find out as much about the drug as we can, and research will help. Dr. Birch needs to continue his research so that he can have a randomized control group in his new studies. Dr. Birch isn't sure yet how well Andrevil is working; some kids could be getting sicker because they are allergic to the drug.

The crux of this argument is that even recognizing that there is some strong counterpositive evidence (C) and risk (M), there may be positive benefits associated with the drug (A) that can only be observed with further research (G) conducted over the long run and/or with larger samples (B, D). Future studies are being planned that have better controls (F), and these are needed before more conclusive results can be obtained (D). The children getting sicker while taking the drug may be affected by other causes, such as drug allergies or natural progression of disease (E).

For the two preceding examples, it may be determined that each argument would have received a total score of 13.

John's Trial. Similar examples can be generated for John's trial. The following is a composite justification for agreeing that boys with an extra Y chromosome should be sent to a special school:

> Because the extra Y chromosome is linked to violence, some Y chromosome people may be dangerous and need to be helped. Even though not all boys with an extra Y will be violent, the special school might help those that will. Also, maybe by studying the people in the school scientists can find out more about the violence and the Y chromosome. It is better to train all boys to be nonviolent at an early age than to wait to see if some will become violent.

This argument is based on belief in the possible existence of a positive relationship between the Y chromosome and violence (A) plus the awareness that there is an acceptable cost/benefit trade-off (M) in the sense that some nonviolent people will be selected (B, C) to go to the school. Argument is stronger if it is acknowledged that there are weaknesses in the existing research (F) and there is awareness that the relationship between Y and violence may not be a real one (D) and that Y does not cause violence (E). The crux of the argument is that despite these drawbacks, there is still willingness to risk selecting boys who are not violent in order to protect society from possibly violent ones. There also seems to be recognition that conducting additional research will help clarify the link between Y and violence (G). Again, the total score for this argument is 13.

An argument for disagreeing with the need for a special school might be:

> Even though some boys may be more violent because of the extra Y chromosome, some boys with Y may not be violent at all. The studies about the relationship are too inconclusive to make decisions by—the evidence for the relationship is not solid. More research is needed. The current research isn't finished and the sample size of the research cited at the trial is too small. Also, there may be risks associated with the special school treatment, as putting kids in such schools might make them angry or feel like outcasts, which could increase the violent tendencies. Also, the testing and special school would cost the taxpayers too much money.

This argument is based on awareness of counterpositive evidence (C) (nonviolent Ys) accompanied by the belief that research showing a relationship between Y and violence (A) is not strong enough to support decision making (F). Because studies showing that correlation are few and poorly designed, the observed relationships may be small in magnitude or due to chance (B, D). Also, even if a correlation exists, Y may not be the causal factor (E). Also, there may be risks associated with school treatment (M): For example, boys may feel outcast or angry and increase their violent tendencies. Also, it may cost taxpayers too much and infringe on freedoms, and so on (M). More research should be conducted to justify any decision making (G). A total score of 13 is once again associated with this argument.

It is shown, then, that although the preceding arguments represent different conclusions drawn from different stories, they would all be scored as equally valid (and 'correct') in terms of our scoring protocol.

DID STUDENTS' THINKING ABILITIES IMPROVE?

Students' abilities to reason statistically did appear to grow as a result of participation in the 3-week activities set. This was shown by a classroom-based analysis (Levin, 1992) of adjusted posttest performance. The summary data are reported in Table 7.2; it was found that the mean statistical-reasoning score of the two experimental classrooms (3.61) was statistically higher than that of the eight comparison classrooms (2.27). Moreover, no performance differences were associated with group versus individual testing.

As classroom research, these results must be interpreted cautiously for a number of reasons. Primarily, the two classrooms that participated in the instructional program did not receive it through the process of random assignment to experimental and control conditions; rather, the experimental classrooms belonged to 2 teachers (out of 10) who volunteered to participate. Thus, the possibility existed that students and/or teachers in the two experimental classrooms initially differed from those in the comparison classrooms.

TABLE 7.2
Classroom-Level Summary Data on the Posttest,
By Condition, Testing Format, and Story Version

	Individual/ Andrevil	Individual/ John	Group/ Andrevil	Group/ John
Adjusted posttest				
Experimental	4.99	2.07	4.33	3.06
Control	3.15	1.51	3.05	1.37
Adjusted percentage of inappropriate responses				
Experimental	18.4%	38.9%	6.4%	42.7%
Control	22.4%	43.8%	40.4%	64.0%

In addition, some data were missing or questionable due to noncompliance of a few students and administrative error, so scores for some students and one entire comparison classroom had to be dropped from our analyses. To the extent possible, we adjusted for these problems in our statistical analyses. Moreover, two preliminary analyses were conducted that showed that experimental and comparison classrooms did not differ statistically with respect to relevant student pretest characteristics.

Although performance gains were evident with both assessment tasks, greater student gains were seen for the Andrevil Trial task than for John's Trial. Thus, whereas both tasks proved to be adequate as assessment instruments, it may be that the Andrevil drug trial was conceptually closer to the actual instructional focus that occurred in the classes.

A number of supplementary analyses were conducted, including a contrast of experimental and comparison classrooms with respect to the proportion of inappropriate explanations included in their responses (see Table 7.2). Experimental classrooms were found to produce a statistically smaller percentage of inappropriate responses (27%) relative to the seven comparison classrooms (43%). Another supplementary analysis examined experimental and comparison students' responses to questions for which statistical reasoning was not appropriate. Recall that on both pre- and postassessment tasks, two randomly ordered questions were asked about the trial scenario that students had read. One of these questions solicited an opinion about a policy issue for which a statistically based response would not be appropriate (e.g., "Should researchers be allowed to receive money from drug companies for their research?").

At one level, students clearly could already discriminate when statistical reasoning was and was not appropriate: Virtually no such inappropriate responses occurred for the John's Trial scenario in either comparison or experimental classes. A few inappropriate statistical responses did occur for the Andrevil Task, but approximately the same number of these errors were made by both experimental and comparison classes. This finding shows that the instructional program produced an advantage for those measures on which an advantage should be expected and desired. That is, instructed students did increase their tendency to use statistical reasoning in situations where such reasoning was appropriate, but did not indiscriminately employ statistical reasoning in situations where such reasoning would not be desired.

DISCUSSION

This study represents a first step in a projected program of research investigating authentic instruction and assessment designed to promote development of "good thinking" within a middle-school student population (and ultimately, society in general). The normative model for good thinking promoted in our instructional program and assessment methodology was gleaned from the literature on statistical reasoning and evidential argument (e.g., Konold, 1989; Kuhn, 1991; Tversky & Kahneman, 1974). It also reflected a reaction to many of the issues surrounding authentic assessment in reformed classrooms discussed in Romberg (1995).

Our view of the purpose of instruction can be summarized as follows:

- Good thinking is like good evidential argument.
- Good evidential argument, in many scientific and everyday contexts, requires statistical reasoning.
- Statistical reasoning does not necessitate computation, but always involves interpreting and reasoning about real-world problems with conceptual structures representing ideas such as probability, correlation, and experimental control.
- An integrated view of the world that incorporates such concepts constitutes a knowledge system for statistical reasoning.
- An important desired outcome of schooling is development of students' knowledge systems for statistical reasoning, including their ability and tendency to activate and appropriately use such knowledge systems in reasoning about societal issues.

Viewed from currently popular streams of research and theory pertaining to instructional methodology, our study represents a quasi-experimental test of instructional design based on constructivist (e.g., Derry, 1996) and situated cognition theories (e.g., Derry & Lesgold, 1996), an approach sharing features with those currently practiced and investigated by European activity theorists (e.g., Hedegaard & Sigersted, 1992; Lompscher, 1989). The following outlines key features of our instructional methodology, which are not tied specifically to the statistical reasoning subject domain:

- Instruction and assessment are grounded in concepts derived from an underlying cognitive-theoretical knowledge system associated with a normative model of "good thinking" within a(ny) problem domain (including interdisciplinary problem domains).
- Direct instruction is limited to presentations that attempt to communicate to students the essence and interrelationships among concepts associated with the normative knowledge system.
- Within students, the construction, integration, and elaboration of the normative knowledge system takes place in the context of goal-directed problem solving.
- Goal-directed problem activity is designed to simulate real-world problem solving, improving the chances that the knowledge system developed during classroom activity will be relevant and transferable to life outside of class.
- Intensive mental construction during social problem solving is encouraged through conversation and argument.
- As conversation and argument occur, thinking activity should be intensively mentored according to cognitive apprenticeship, which involves performance modeling, scaffolding, and fading by instructors (e.g., Collins et al., 1989). Mentoring is intended to influence the form and structure of the knowledge systems that students develop.

Implementation of our program here was a preliminary adventure and did not meet all of these ideals. As described in another chapter detailing our subjective observations of this program in action (Osana, Derry, & Levin, 1996), this intervention, particularly the culminating simulation activity, unfolded in sometimes unexpected ways. To some extent we felt unsatisfied with our direct model presentations and with the uneven and unstructured quality of mentorship that the students received. Discipline, student attitudes, and the physical arrangement of the classrooms were far from ideal, and these problems were greater in one experimental classroom than in the other. Nevertheless, when their arguments were judged against a normative model of good thinking grounded in cognitive research on statistical reasoning and evidential argument, students in two classes that participated in our 3-week instructional program performed statistically better than students in comparison classes that did not participate. The students who participated in the instructional program did not overgeneralize what they had learned. They apparently were able to discriminate between contexts in which statistical reasoning was appropriate and contexts in which it was not. Interestingly, however, there was no evidence that students tested in groups were able to reason better than those who were tested individually, suggesting that improved reasoning scores were not dependent on the presence of social support during problem solving.

A useful by-product of this work has been development of assessment tasks and, more importantly, a scheme for gauging the reasoning in students' responses that might be used over a broad range of tasks and student ages (we have now used it at both the eighth-grade and the college level). Importantly, our instructional intervention in no sense "taught to the test," and yet our assessment procedure was sensitive to important changes in students' reasoning ability that should occur as a result of statistics instruction. The interesting finding in this case is that students' abilities appeared to develop as a result of a classroom intervention that was strongly oriented toward learning from mentored activity rather than from direct lecture-based instruction.

As previously argued, we cannot draw definite causal conclusions from these results, primarily due to lack of random assignment of classrooms to instructional conditions. We maintain, however, that the results strongly suggest that our instructional intervention did support growth in statistical reasoning capabilities. Further refinements in the instructional units, the simulation activities, the implementation procedure, and the assessment are clearly needed, and should serve to enhance the impact of such instruction. Although more definitive conclusions must await further research, our work to date provides encouragement that students' reasoning abilities can be enhanced through instructional programming that is grounded in statistical reasoning concepts and that emulates the mentored, contextualized, social activity characteristic of many natural learning environments found outside of school.

ACKNOWLEDGMENTS

The authors acknowledge significant contributions by Velma Hamilton Middle School teachers Tom Bauer, Kim Vergeront, Doris Dubielzig, and Joan Ammacht. We also thank Leona Schauble for her assistance with instruction and for many valuable suggestions throughout the project. Jasmina Milinkovic is thanked for help in planning and data collection. We are also grateful to Ian Duncan, of the University of Wisconsin–Madison's Department of Veterinary Medicine, for his presentation related to the film *Lorenzo's Oil*. Finally, we are grateful to the National Center for Research in Mathematical Sciences Education (NCRMSE) for its financial support.

REFERENCES

Collins, A., Brown, J. S., & Newman, S. E. (1989). Cognitive apprenticeship: Teaching the craft of reading, writing, and mathematics. In L. Resnick (Ed.), *Knowing, learning and instruction: Essays in honor of Robert Glaser* (pp. 453–494). Hillsdale, NJ: Lawrence Erlbaum Associates.

Derry, S. (1996). Cognitive schema theory in the constructivist debate. *Educational Psychologist, 31*(3/4), 163–174.

Derry, S., & Lesgold, A. (1996). Instructional design: Toward a situated social practice model. In D. C. Berliner & R. C. Calfee (Eds.), *Handbook of educational psychology* (pp. 787–806). New York: Simon & Schuster Macmillan.

Derry, S., Levin, J. R., & Schauble, L. (1995). Stimulating statistical thinking through situated simulations. *Teaching of Psychology, 22,* 51–57.

Dewey, J. (1938). *Experience and education.* New York: Collier.

Hedegaard, J., & Sigersted, G. (1992). Experimental classroom teaching in history and anthropological geography. *Multidisciplinary Newsletter for Activity Theory, 11*(12), 13–27.

Konold, C. (1989). Informal conceptions of probability. *Cognition and Instruction, 6,* 59–98.

Kuhn, D. (1989). Children and adults as intuitive scientists. *Psychological Review, 96,* 674–689.

Kuhn, D. (1991). *The skills of argument.* Cambridge, MA: Cambridge University Press.

Lave, J. (1991). Situating learning in communities of practice. In L. Resnick, J. Levine, & S. Teasley (Eds.), *Perspectives on socially shared cognition* (pp. 63–82). Washington, DC: APA Press.

Levin, J. R. (1992). On research in classrooms. *Mid-Western Educational Researcher, 5,* 2–6, 16.

Lompscher, J. (1989). Formation of learning activity in pupils. *Learning and Instruction, 2.2,* 47–66.

Nisbett, R. E., Fong, G. T., Lehman, D. R., & Cheng, P. W. (1987). Teaching reasoning. *Science, 198,* 625–631.

Osana, H. P., Derry, S., & Levin, J. R. (1996). Developing statistical reasoning through simulation gaming in middle schools: The case of "The Vitamin Wars." In B. G. Wilson (Ed.), *Constructivist learning environments: Case studies in instructional design* (pp. 83–92). Englewood Cliffs, NJ: Educational Technology Publications.

Romberg, T. A. (Ed.). (1995). *Reform in school mathematics and authentic assessment.* Albany, NY: SUNY Press.

Shaughnessy, J. M. (1992). Research in probability and statistics: Reflections and directions. In D. A. Grouws (Ed.), *Handbook of research on mathematics teaching and learning* (pp. 465–494). Reston, VA: National Council of Teachers of Mathematics.

Tversky, A., & Kahneman, D. (1974). Judgment under uncertainty: Heuristics and biases. *Science,* 185, 1124–1131.

von Glasersfeld, E. (1990). An exposition of constructivism: Why some like it radical. *Journal of Research on Mathematics Education,* Monograph No. 4. Reston, VA: National Council of Teachers of Mathematics.

Vygotsky, L. S. (1978). *Mind in society: The development of higher-psychological processes* (M. Cole, V. John-Steiner, S. Scribner, & E. Souberman, Eds. and Trans.) Cambridge, MA: Harvard University Press.

APPENDIX A
Lorenzo's Oil Group Discussion Guide
Day 3

Date _____   Teacher _____   Group_____

Facilitator (name) _____

Recorder (name) _____

Lorenzo's parents did their research by trying to learn about and understand what *caused* Lorenzo's illness. The medical researchers were trying out treatments on many children to see what worked and what did not. What differences do you see in the two types of research? Is one type better than the other? Do we need both?

1. The main difference between the research of the medical scientists and that of Lorenzo's parents was:

_____

_____

_____

_____

2. Do you think one type of research is better than the other, or do we need both? Explain your answer.

_____

_____

_____

_____

_____

_____

_____

APPENDIX B
READ THE FOLLOWING PAGES VERY CAREFULLY.
YOU MAY WRITE ANYWHERE YOU LIKE.
The Andrevil Trial

I. INTRODUCTION

Mrs. Wilson's son Kenny has been taking an experimental drug for a serious disease. Mrs. Wilson believes that the drug made Kenny sicker than he was. She and Kenny are suing Dr. Birch, the scientist who gave Kenny the drug. They are trying to stop Dr. Birch from doing more studies and giving the drug to other children.

In the scene you are about to read, there are two lawyers and two witnesses. One lawyer is on Kenny's side. The other lawyer represents Dr. Birch. The first witness is Kenny's mom. The second witness is Dr. Birch.

II. COURT SCENE

A. First witness questioned by Kenny's lawyer.

Kenny's Lawyer (KL): Mrs. Wilson, are you familiar with the drug called Andrevil?
Kenny's Mom (KM): Yes, it is a new drug for the treatment of a severe form of cancer.
KL: How do you know about this drug?
KM: My son Kenny has this severe form of cancer. Last year the University did a study on Andrevil. Kenny was in it and took Andrevil for several weeks.
KL: Did Kenny improve?
KM: No, he became very ill. He got worse. He is now much sicker than he ever was.
KL: Was Kenny the only one who became sicker?
KM: No. I know other parents with children like Kenny. Two of their children who were in the study became very ill. One of them died.
KL: When Kenny became ill, did you notify the doctor in charge of the study?
KM: Yes. That would be Dr. Birch. He said that some children were expected to become sicker, but that the drug still might be working.

B. First witness questioned by Birch's lawyer.

Birch's Lawyer (BL): Mrs. Wilson, why did you put Kenny in the Andrevil study in the first place?
Kenny's Mom (KM): He was very ill. We felt it was his last chance.
BL: When you put Kenny in the study, were you promised that the Andrevil would improve his condition?
KM: No, but . . .
BL: When you contacted Dr. Birch to say that Kenny was getting worse, did he remind you that you could remove Kenny from the study if you felt it was doing him harm?

KM: Yes, but he advised against it.
BL: Why did you keep Kenny in the study if he was getting sicker?
KM: Because Dr. Birch said the drug might make some children sick, but it should still work. He said the theory behind it was good.

C. Second witness questioned by Kenny's lawyer.

Kenny's Lawyer (KL): Dr. Birch, how many children in your study took Andrevil?
Birch: Nineteen children.
KL: How many of these children have now died?
Birch: Four.
KL: And of those still living, how many are improved and how many are worse?
Birch: Four or five have improved, but four or five are now worse.
KL: So nearly half the children who took Andrevil are now either dead or sicker than they were?
Birch: Yes, sadly, I'm afraid that is true.
KL: But despite these deaths and sicknesses, you want to continue your research with Andrevil?
Birch: Yes, we must.
KL: Dr. Birch, would stopping your research with Andrevil damage your career?
Birch: It would not help it. Most of my research money is for Andrevil studies.

D. Second witness questioned by Birch's lawyer.

Birch's Lawyer (BL): Dr. Birch, please explain why you believe research with Andrevil should continue.
Birch: We know that the children in our study were all in late stages of their disease, without much hope. Some of these children are still alive and feeling good today. This might be due to the Andrevil. We must find out.
BL: How will you find out?
Birch: We will watch the Andrevil group for a number of years. We will also watch a group of sick children who did not receive Andrevil. If more of the Andrevil children live longer and better lives, then this is evidence that the Andrevil is working.
BL: But why give more Andrevil to more children if you can get the answer from this one study?
Birch: One small study will not give us the answer.

Here are some questions to think about:
1. Mrs. Wilson wants to stop Dr. Birch's research with Andrevil. Dr. Birch believes his research must continue. Do you agree with Mrs. Wilson or Dr. Birch?
WHY?
2. Should researchers like Dr. Birch be allowed to receive money from drug companies for their research?
WHY?

PART FOUR

ASSESSING STATISTICS

CHAPTER EIGHT

Monitoring Student Progress in Statistics

Susanne P. Lajoie
Nancy C. Lavigne
Steven D. Munsie
Tara V. Wilkie
McGill University

Statistics: The very word has been known to induce pallor in a graduate student. Such distress over statistics arises for many reasons, some of which include the abstract nature of some of the concepts, the mathematical treatment of the subject matter in many statistics classes (e.g., emphasis on computation), and students' insufficient background prior to entering such classes. These factors are of prime concern in many research endeavors, as seen in some of the chapters in this volume. One suggestion for alleviating this distress for future students is to provide early instruction of statistics, spread over time and across grade levels. But how do we make the subject matter that graduate students want to avoid appear relevant to younger students who do not see the utility of mathematics in their daily lives? Therein lies the challenge that motivates the research described in this chapter.

The goal of this research, referred to as the Authentic Statistics Project (ASP), has been to make statistics meaningful to students in middle school, grade 8 in particular, and to assess the type of progress students achieve in learning statistics. One way of enhancing the value of statistics to middle-school students is to demonstrate how statistics can be used to answer important questions and make everyday decisions. Within this context, students can learn to perceive statistics as a valuable tool rather than a needless subject matter.

In the ASP, students use statistics to answer questions that they find meaningful. Their activities focus on investigations that are self-generated whereby statistics are used to answer their own questions, making the inquiry process meaningful. To create the ASP environment, a mathematics classroom was transformed into small groups of design teams, where each team constructs a research question, collects data to answer it, represents the data graphically, analyzes the data, and presents its project to the classroom. Each group is provided with technological supports that are intended to facilitate the investigation process. Technology serves to facilitate both instruction and assessment. We demonstrate the ways in which student progress in statistics can be monitored in this form of classroom instruction. A more complete description of the ASP follows, with pointers to its theoretical foundation, as well as an in-depth examination of two case studies from ASP.

THE AUTHENTIC STATISTICS PROJECT

The ASP has a problem-solving focus where students learn to apply statistics to their everyday lives by engaging in the process of investigation. Opportunities for students to engage in "doing" statistics are provided. Doing statistics then reflects sense-making (Schoenfeld, 1985), thinking skills, and constructions of knowledge, rather than the learning of algorithms (Cognition and Technology Group at Vanderbilt [CTGV], 1992; Resnick, 1987).

The ASP classroom is set up with eight computer workstations (Macintosh), and groups of two to four students work at each station to learn descriptive statistics. A classroom teacher monitors the students' work but is assisted by tutors (graduate students) who work with the individual groups. Students' statistical understanding is assessed prior to, during, and after the ASP has been completed. For ease of discussion, we have isolated five stages of the ASP that are identified as time 1 through time 5 in Fig. 8.1. The criteria used to assess students' performance on statistics tasks are also identified in this figure. At time 1, students are administered a pretest examining their knowledge of statistics prior to their participation in our research. Time 2 represents the 4-day tutorial where instruction focuses on teaching statistical content as it relates to the statistical investigation process. Statistics are taught in situ; that is to say, statistics are taught through concrete examples where students must use computer software to graph and analyze data. Time 3 follows this knowledge acquisition phase with concrete models of statistical performance. The models are presented in a library of exemplars that models performance standards for designing and engaging in statistical investigation. During time 4 students apply what they have learned in the instruction and modeling phase to the design of their own investigations/projects. Posttests are administered at the end of the study at time 5. Brief descriptions of the tutorial, library of exemplars, and the design phase are discussed next.

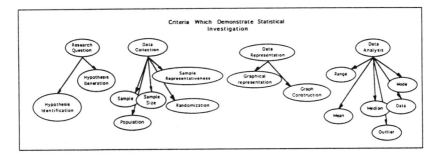

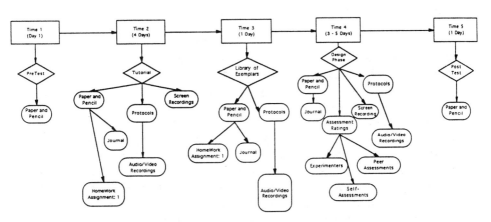

FIG. 8.1. Assessment criteria of statistical understanding over each time stamp in ASP.

The Tutorial

The major objective of our research was to provide instruction that enabled students to learn statistical content and apply such content in a variety of problems, particularly within the framework of inquiry or investigation. A tutorial was designed to meet this objective. Students were taught descriptive statistics, such as measures of central tendency—mean, median, and mode (see Fig. 8.1)—hand-in-hand with procedures for "doing" statistics through the use of computer software. The tutorial, for example, required that students use MyStat, a statistical package, and CricketGraph, a graphics package, to learn about statistics concepts in terms of data analysis and representation procedures. This instruction was embedded within the overall investigation process. In this sense, students were expected to acquire and apply both statistical knowledge and computer skills in an applied and meaningful context.

The tutorial provided students with four activities that enabled them to gain and utilize statistical knowledge and computer skills. Two of these activities, the pulse rate and school grades activities, provided students with opportunities to learn concepts in terms of statistical procedures through the use of computer tools. Concepts such as the mean and range were taught in terms of data analysis procedures that were used to answer a particular question. For example, in the pulse rate activity,[1] students collect and analyze pulse rate data by taking each individual pulse rate, entering the data into the computer, and computing the group mean. Each group collected at-rest pulse rates in the group, entered its data into the computer, and analyzed it using MyStat. Individual group data were then compared to data from the whole class. Discussions about the mean, mode, and median, data, sample, population, and range were generated in the context of these activities. Following this exercise another group of students "ran on the spot" for a minute and students collected, analyzed and interpreted their data by comparing it with the "at-rest" data. By comparing two group means predictions were made, hypotheses were tested, and results were interpreted. These activities required whole-class participation. Students were coached by instructions in the tutorial and by the teachers in how to use the computer to enter, analyze, and graph data.

The "school grades activity" helps students examine the influence of extreme scores and outliers on the mean grade. A sample file of grade 8 mathematics scores was provided. The first exercise introduced students to the notion of extreme scores and how they affect the mean. The raw data consisted of: 50, 52, 56, 94, 98, 100. They were asked to enter these numbers into the computer and analyze the data. The mean grade was 75. Students were then asked whether or not any students received a grade of 75 or close to it. Given that this was not the case, students were introduced to the notion of extreme scores and bimodal distributions and how means are not always representative of the average person in the sample. The pulse rate and grade activities build on each other and allow further exploration of statistical content such as sampling (i.e., population, sample, the size and representativeness of a sample, and randomization). Two additional activities are included in the tutorial to reinforce this learning, but in terms of the interpretation and use of graphical representations for making predictions about data.

The intent of the tutorial was to provide students with the necessary knowledge and skills that could be used in their own statistics projects. Our goal was to examine how students apply the statistical and software knowledge acquired in the tutorial to the investigation process. As such, students were expected to design a statistics project that required engaging in statistical investigation, that is, the design phase of our study. However, we

---

[1]This activity was a modification of an American Statistical Association (1991) activity designed for K–12 mathematics.

believed that we needed to give further examples of statistics in use in the context of statistical investigations before students could feel confident in meeting our expectations. In an effort to make the teacher expectations and assessment goals clear, students were given models or exemplars prior to designing their own statistics projects. A technological tool, the library of exemplars, was used to demonstrate these goals, and to make assessment criteria clear. This tool, the library of exemplars, is described next.

The Library of Exemplars

The library of exemplars is a computer tool designed to provide students with an instructionally relevant and informative learning environment that enables students to reflect on the learning process. Such reflection can be facilitated by making the instruction and assessment goals clear to students. This objective can be accomplished by making the assessment standards open or transparent to students prior to a performance (National Council of Teachers of Mathematics [NCTM], 1995). Such transparency ensures that those who are being assessed understand the criteria on which they are being judged so they can improve their performance (Frederiksen & Collins, 1989; Wiggins, 1992). Making scoring criteria open through genuine benchmarks, that is, examples of student work that meet criteria rather than arbitrary cut scores such as achieving 60% on a test, can make the task more comprehensible, thereby allowing a learner to demonstrate what he or she truly knows.

In an effort to help students reflect on how the knowledge they acquired in the tutorial can be applied to a performance task involving statistical investigation, a library of exemplars was created. The library of exemplars is presented to students after the tutorial and prior to the design of their own statistics projects. Although the tutorial provided students with opportunities to engage in various aspects of the statistical investigation process (e.g., pulserate activity) it was still heavily factually oriented. The library of exemplars was more explicit in its focus on the components of this process, so that students can strive to accomplish these benchmarks in their own projects. The components consist of (a) generating a research question, (b) collecting data to address this question, (c) analyzing the data, and (d) representing the data (Lajoie, Jacobs, & Lavigne, 1995; Lavigne & Lajoie, 1996). The library of exemplars illustrates how students can perform on each of these components in a statistics project by providing two exemplars, "average" and "above average," of how previous groups of students designed and conducted such a project in terms of each component.

Because the library was designed to make the instruction and assessment goals clear to students, each component of the statistical investigation process was represented as an assessment criterion. For each criterion, the digitized example was presented with a textual description of the component, how it was scored, and questions to facilitate discussion among the group about how the two examples differed. For example, students examining the research question criterion were given two examples of students who designed a research question. One group clearly specified the categories in the question and was given a maximum score of 5 points. Its question was, "What is your favorite fast food restaurant between Harvey's, Burger King, McDonald's, LaFleur's, and other." Another example, which received 3 points, simply stated, "What is your favorite fast food restaurant?" Students were asked to discuss which example was a better research question and why. Viewing performance standards in this way was expected to help students reflect on the strengths and weaknesses (Nitko, 1989) of other student projects, set goals for their own group projects, and monitor their progress toward achieving those goals. The library of exemplars was thus expected to facilitate students' performance in the design phase of ASP where they had to plan and conduct their own statistics projects. This design phase is described next.

The Design Phase

During the design phase of ASP, students apply what they have learned in the tutorial and library of exemplars to their own statistics projects, which they design in collaborative groups. These projects require that group members engage in many activities such as communicating their ideas (orally or in writing), listening to the ideas of others, making decisions about the appropriateness of a research plan, collecting the data, entering the data, making decisions regarding how to graph and analyze the data, and finally interpreting the data. In this sense, the ASP is a project-based approach to instruction.

Project-based instruction, in general, is a comprehensive approach to classroom teaching and learning. This approach is designed to engage students in complex investigations of realistic problems that integrate concepts from a number of disciplines and extend over a long period of time (Blumenfeld et al., 1991; DuCharme, 1993; Krajcik, Blumenfeld, Marx, & Soloway, 1994; Scott, 1994; Trepanier-Street, 1993). ASP fits this description in that statistics is taught from a problem-solving perspective through an extended group project. Tasks that are presented in meaningful settings serve as connectors or bridges between classroom instruction and real-world contexts. Students seek solutions to nontrivial problems by asking and refining questions, debating ideas, making predictions, designing plans and/or experiments, collecting and analyzing data, drawing conclusions, and communicating ideas and findings to others. Through such inquiry, a learner may be required to question existing ideas and perhaps be forced to reorganize his or her existing schemas. In addition to learning, motivation may increase. Blumenfeld et al. (1991) found that cognitive engagement increases when students are responsible for the creation of questions and activities. The generation of questions is critical because students are constructing new knowledge in a dynamic way, in which the doing and learning are inextricable.

The project approach is not new (Blumenfeld et al., 1991; Trepanier-Street, 1993); it has been used primarily in early childhood education (DuCharme, 1993; Gorb, 1987) and more recently in middle school (Blumenfeld, Krajcik, Marx, & Soloway, 1994; Edgerton, 1993; Krajcik et al., 1994; Ladewski, Krajcik, & Harvey, 1994; Scott, 1994; Sommers, 1992; Wolk, 1994). It also shares some common features with the anchored instruction approach (CTGV, 1992; Schwartz, Goldman, Vye, Barron, & CTGV, chapter 9, this volume) where problem-solving environments have been designed to provide "macrocontexts." Macrocontexts involve complex situations that require students to formulate and solve a set of interconnected subproblems (Bransford et al., 1988).

The theoretical justification for this project approach is based on our view that learners are active problem solvers rather than passive recipients of knowledge. Obviously, this is not a new insight. The notion of the learner as an active participant has roots in Kant's (1787/1968) philosophy of "learning by doing." However, in ASP, we are starting to examine the nature of "doing" in realistic and situated contexts, in that students are selecting projects that are meaningful to them and the learning of statistics is situated in these contexts. Because this process is done in groups, the role of the learner in a community of learners must also be addressed. Social relationships are a means of internalizing situations and creating consensual domains (Vygotsky, 1978). Such learning communities can facilitate meaning construction (social constructivist theory—von Glasersfeld, 1988) and promote what Gal (chapter 10, this volume) refers to as a culture of explaining.

Guidance for providing appropriate social facilitation through group project activities is presented in cognitive apprenticeship models (Collins, Brown, & Newman, 1989). We have adapted this model in ASP by ensuring that multiple contexts for learning statistics are provided throughout the instructional process, as well as providing a culture of expert practice (Lajoie et al., 1995). The community of learners provides some of these contexts, especially in the design phase, where students share ideas for planning and conducting their investigations. However, these contexts are extended in the formal presentation of projects by each group. Hence, multiple opportunities are provided for extending student understanding of the statistical investigation process through the dialogues between students, as well as between students and teachers and other mentors. A culture of expert practice is provided because teachers, tutors, and more skilled students help students to reason about and with statistics during every phase in ASP. These experiences are expected to help each group assess itself subsequent to the presentation, in terms of the performance standards for the four statistical components observed in the library of exemplars (see Fig. 8.1). The class also evaluates each of the other groups on these criteria, as do the teachers and tutors. These multiple ratings provide the teachers and researchers with evidence as to whether or not the students have interpreted the performance standards in the same manner, and whether or not there is agreement among the raters. The student-generated statistics projects serve as excellent assessment tools, because they are the final products of the instructional process. The project work encapsulates what students have learned over time and how they applied such knowledge to their statistical investigations. The rationale for monitoring this type of student progress is described in the following section.

MONITORING STUDENT PROGRESS

What Is the Purpose of Monitoring Student Progress?

The *Assessment Standards* (NCTM, 1995) suggest that one of the primary purposes of assessment is to monitor student progress in mathematics. Monitoring student progress refers to judging progress toward attaining mathematical goals or, in this chapter, statistical goals. The general goals of the ASP were that students learn to problem solve, reason, and communicate their understanding of statistics. These general curriculum goals are consistent with the NCTM *Curriculum and Evaluation Standards* (1989). Because the ASP required that students learn and employ these skills in the context of statistical investigation, our documentation of student growth focuses specifically on the components of the statistical investigation process rather than on general curriculum goals.

One purpose of monitoring student progress is to communicate to students about their performance (NCTM, 1995). By providing continuous monitoring in the context of assessment, it is possible to provide the appropriate level of feedback to facilitate students' progress in meeting their objectives. Communicating with students about their performance helps them to become aware of the goals of instruction and assessment and to learn to assess their own work. Documenting student progress is best done through the use of multiple assessment tools, because there is a greater likelihood of capturing what students know through multiple forms of assessment than through single snapshots of the learner.

One of the complications in describing the impact of the ASP work is finding ways to gather, synthesize, and report the evidence we have collected to communicate the richness of the instructional experience. To this end, we provide two case studies, Magic and Metallica, to demonstrate how different forms of assessment can be used to provide meaningful learning profiles. The case study approach draws from a variety of field-based methods borrowed from sociology, anthropology, and political science (Romberg, 1992). Our evidence is gathered through observations, pre and post paper-and-pencil tasks, structured journals, verbal protocols of group interactions and presentations, and the examination of artifacts, such as those created on the computer to be used in the project presentations (screen recordings). Two criteria were applied to the selection of Magic and Metallica: (a) the completeness of the data set and (b) the quality of the final group presentations. Obviously, the more complete the data, the better the argument that can be made about each group's statistical understanding. We selected groups that varied somewhat in their final presentations; one group excelled and one group did moderately well. Our goal is to tell a story about the strengths and weaknesses of each group and to compare the groups on features of statistical understanding. In so doing, we present an argument about the patterns that define the range of findings (Romberg, 1992) that can be expected in this work. It has been suggested that vignettes or stories may be a more effective form of communicating results to practitioners, thereby enhancing the implementation of new methods of instruction and assessment in classrooms (Berliner, 1992; Carter, 1993). We have tried, to the extent possible, to tell such stories.

What to Monitor

As illustrated in Fig. 8.1, many methods of assessment were developed to monitor student progress on the performance standards that best reflect the statistical investigation process. We were particularly interested in tracking how students progressed on these components over time. For some components, we were able to track in all five time slots; for other components we had more limited data. We monitored students' understanding of a research question in two ways, one where they had to identify what the hypothesis was in a given scenario and another where students designed research questions. When students collected data to complete their projects, we were most interested in how they applied their statistical understanding to the data collection process. For instance, did they understand the concepts of sample, sample size, sample representativeness, population, and randomization? Students' progress in acquiring data analysis skills was examined in terms of how well they interpreted data, and how well they demonstrated their understanding of key concepts such as range, mean, median, mode, data, and outlier. When students prepared their data for class presentations, we monitored them on the types of graphs they constructed and how well they interpreted their graphical representations. Having explained what is monitored, we next describe how to monitor.

How to Monitor Student Progress in Statistics?

In an effort to judge student progress toward achieving statistical goals, multiple forms of assessment are needed: that is, performance tasks, projects, writing assignments, oral demonstrations, and portfolios (NCTM, 1995). In addition to examining the type of evidence provided through one assessment medium, it is important to examine multiple forms of evidence jointly to see whether or not one medium can elaborate our understanding of the learner.

Each medium of assessment in ASP may be used to indicate a different type of learning or provide for an alternate manner for monitoring student progress. For example, paper-and-pencil measures can be in the form of quizzes or structured journals. Each measure has different strengths. A quiz can provide a summary of basic knowledge, whereas a structured journal may go beyond a summary by allowing students opportunities to explain their work through embedded prompts. Audio- and videotapes of students as they work on their statistics projects provide the teacher or assessor with opportunities to review what they may not have had time to do in class. These tapes often provide insights into the reasoning processes of students in the context of their problem solving activities. Computer screen recordings provide concrete examples of student work. A computer screen recording is an exact recording of every action a student conducted on the computer in the context of their projects using CricketGraph and MyStat. The recording can be replayed just like a videotape. By analyzing the screen recordings it is possible to critically review the number and types of graphic manipulations made (i.e., pie charts, bar charts) along with the types of statistical manipulations conducted (i.e., frequencies, means). These recordings provide direct evidence of how students interact with the raw data they have collected and provide opportunities for detecting student misconceptions or difficulties encountered using the computer.

Starting with the screen recording data, a teacher might note that students were having difficulty organizing their data for subsequent analysis, but that over time these same students had become proficient at data organization. The benefits of repeated assessments are that such changes in students' understanding can be noted. Problems encountered early in the knowledge acquisition phase may disappear by the end of the design process. Adding multiple forms of evidence to our understanding of this group can be done by analyzing the audiotapes of this group that were taken when the screen recordings were made, allowing us to develop a better picture of their reasoning processes while they use the computer to develop their statistics projects. The group's responses to pre and post paper-and-pencil tests, homework assignments, and journals can also be used to build a more robust picture of the group's statistical understanding. For example, do members of the group elaborate on the design of their research question in their journal, and do their verbal protocols reveal their understandings or misunderstandings about graphical and statistical manipulations?

Providing Valid Indicators of Learning

What evidence is there that these new forms of assessment provide valid indicators of what has been learned? As stated previously, 2 groups of students (Magic and Metallica) were selected from a sample of 16 groups participating in this study. The groups are compared in terms of their understanding of each statistical investigation component (i.e., research question, data collection, data analysis, and data representation) and associated statistical concepts through multiple mediums of assessment over time. These comparisons are based on scores assigned on point systems devised to assess quality as well as accuracy of understanding. These scores were then converted to percentages, which were used to compare groups.

Case Overview

As stated earlier, the two groups described here were selected based on the variation in their performance throughout ASP as well as the completeness of data collected from them. Both groups were from the same grade 8 classroom in a public high school in an anglophone suburb of Montreal. Each group was mixed ability and female, with Magic having two and Metallica having four students. A brief overview follows that describes the groups' overall statistical knowledge, types of statistical knowledge acquired, and evaluation of the ASP experience. This overview is followed by a more in-depth look at each case in terms of the statistical investigation process.

Overall Statistical Knowledge. As mentioned previously, the instructional treatment in ASP consisted of three phases: (a) the knowledge acquisition phase where students learned statistical content and computer skills in a problem-solving tutorial; (b) a library of exemplars phase where concepts were more explicitly embedded within components of statistical investigation and guidelines and expectations about performance on such components were made clear to students; and (c) a design phase where students applied their knowledge and skills in their own project requiring statistical investigation. Each phase was designed to foster and refine students' statistical knowledge.

To determine the course of knowledge acquired by students in each group as a result of participating in ASP, pre- and posttest scores were examined. The pre- and posttests provided students with opportunities to demonstrate the knowledge they acquired from participating in all three phases of the instruction. The test items were consistent with the instruction and required some definitions of concepts and calculations, but focused primarily on explanations of procedures and how they would be used in a particular context. One test item, for example, required that students explain how they would collect data for examining grade 8 students' favorite fast food restaurants in the province of Quebec. This task required that students decide on an appropriate sampling method and sample size, as well as determine how to ensure the representativeness of the sample.

The groups' mean scores on the pre- and posttests, presented in Fig. 8.2, illustrate that Magic and Metallica did not differ in their statistical knowledge

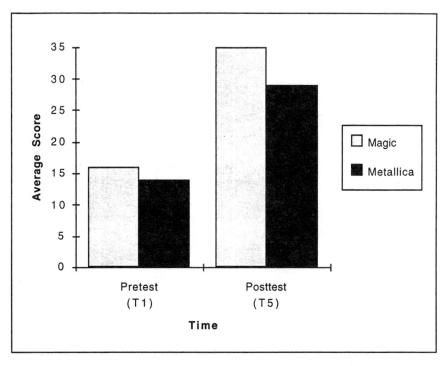

FIG. 8.2. Each group's performance on pre- and posttests.

prior to participating in ASP. However, by the end of our study, both groups'
knowledge increased, with Magic acquiring more knowledge than Metallica.
An analysis of variance [group (Magic, Metallica) × time (pre, post) × accu-
racy] indicated that this group difference is not statisically significant ($F =$
0.795, $p > .05$). Nor was there a group by time interaction ($F = 0.570$, $p >$
.05). However, there was a main effect of time ($F = 22.431$, $p < .001$),
indicating that accuracy increased for both groups from pre- to posttest. This
finding suggests that the two groups gained significant statistical knowledge
as a result of the ASP instruction. However, because we are interested in
the case profiles and in detailed comparisons of performance, additional
sources of evidence are used. In keeping with the case study approach, all
of the remaining data reported next are descriptive.

Types of Knowledge Acquired. Declarative, procedural, and conceptual types of knowledge have been described at length by Anderson (1983)
and are currently being used to describe a statistics curriculum that is embedded in an computer-based learning environment for college level students, called StatLady (Shute & Gawlick-Grendell, 1994). For the purposes
of the ASP work, we have identified each of the assessment measures found
in Fig. 8.1 and Table 8.1 according to the type of knowledge assessed.

TABLE 8.1
Directional Breakdown of Group Differences by Criteria by Knowledge

	Magic Performs Higher Than Metallica			Metallica Performs Better Than Magic		
	Declarative	Procedural	Conceptual	Declarative	Procedural	Conceptual
Research Question						
ResIdentpre				X		
ResIdenttut				X		
ResIdentpost				X		
ResGenpre				X		
ResGenlib			X			
ResGenproj			X			
ResGenpost			X			
Data Collection						
Datacollpre	X		X			
Datacolltut					X	
Datacolllib			X			
Datcollproj	X	X	X			
Datacollpt				X		X
Data Analysis						
Datanalpre	X	X	X			
Datanaltut	X	X				
Datanallib			X	X	X	
Datanalproj		X	X			
Datanalpost	X	X	X			
Data Representation						
Datprespre	X	X	X			
Datprestut		X	X			
Datpreslib						X
Datpresproj		X	X			
Datprespost	X	X	X			

Declarative knowledge refers to factual or symbolic knowledge, such as defining *random sample*. Procedural knowledge refers to how to do something, such as how to calculate the mean based on data found in table x. Conceptual knowledge refers to the ability to apply one's knowledge to new situations and to being able to explain why something is done. For instance, given a scenario on athletes and use of steroids, can students determine, from the given data, whether or not the conclusions drawn from the study were legitimate and generalizable to all athletes? In general, the descriptive data reveals that although both groups acquired statistical knowledge, they differed in the types of knowledge gained. Table 8.1 presents a breakdown of group strengths, describing which group excelled on which statistical components at different assessment times.

The detailed breakdown in Table 8.1 provides us with a better understanding of group profiles. Although the previous analyses indicated that Magic and Metallica performed similarly in their overall post test performance, Table 8.1 illustrates that they differed vastly in their performance on individual group projects and in the type of knowledge they acquired during ASP. The only similarity between the two groups was in their development of surveys as projects. Interestingly enough, this similarity may be due to the library of exemplars, which explicitly modeled how to conduct a survey (favorite fast food restaurants) under the research question criterion (Lajoie, 1996). Magic's survey involved favorite basketball teams among the Chicago Bulls, New York Knicks, Orlando Magic, Portland Trailblazers, Boston Celtics, Pheonix Suns, and San Antonio Spurs, whereas Metallica examined preferences in favorite kinds of music among pop, dance, reggae, heavy metal, or rap. The difference lies in the groups' performance on components of the statistical investigation process. As Table 8.1 shows, Magic clearly outperforms Metallica on all the components, demonstrating better procedural and conceptual knowledge in their performance on projects. Figure 8.3 more clearly illustrates Magic's lead over Metallica in all the categories of investigation. Table 8.1 also indicates that Magic's strength in conceptual knowledge is consistent across all measures. This type of assessment, where knowledge type is identified with the type of task, allows one to have a better understanding of how individuals, or in this case groups, differ and whether or not multiple assessments provide consistent information about learning.

To obtain a better picture of the type of knowledge acquired by each group across all measures, proportions of accurate responses to test and homework items, journal prompts, and questions during presentations of projects specific to statistical investigation components were computed. These proportions, presented in Table 8.2, indicate that Metallica and Magic are similar in their procedural knowlege, but different in their acquisition of declarative and conceptual knowledge despite the same instruction. Also,

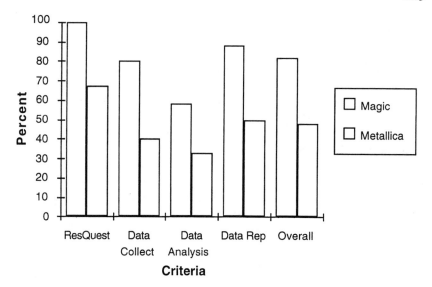

FIG. 8.3. Each group's performance on performance standards (statistics criteria) for statistics projects.

Metallica is much higher than Magic in terms of their strength in declarative knowledge, but Magic surpasses Metallica in terms of their conceptual knowledge.

Evaluation of ASP Experience. Overall, both groups learned from ASP, but we were also interested in whether or not they enjoyed the experience. An evaluation of ASP demonstrated consensus that students enjoyed the ASP experience. When asked what they liked about the course, the groups agreed that they liked using computers to develop new learning skills, working in groups to discover statistics, conducting their own studies, and using graphs to visualize their results. Both groups said they wished that the ASP was longer in duration and that printers were available to print their files.

TABLE 8.2
Proportions of Accurate Responses by Type of Knowledge by Criteria

	Declarative		Procedural		Conceptual	
	Magic	*Metallica*	*Magic*	*Metallica*	*Magic*	*Metallica*
ResQuest	0	100	0	0	100	0
DataColl	33	33	17	33	50	33
DataAnal	27.3	50	36.4	50	36.4	0
DataRep	18	0	36	0	46	100
Overall	20	46	22	21	58	33

Statistical Understanding Based
on Statistical Investigation Components

The following sections further describe Magic and Metallica in terms of the performance standards that reflect components of the statistical investigation process, namely, identifying and generating a research question, data collection, data analysis, and data representation.

Research Question. The groups of students were given two types of tasks related to statistical research questions: (a) identifying relevant properties of a research question presented in a problem, and (b) generating a clear research question for a particular problem and for designing a statistics project. Research identification tasks were more constrained than research generation tasks. An example identification task presents students with a problem in which a researcher is interested in determining whether high-school students learn mathematics better with computers than without computers. Additional information is provided regarding data collection and analysis procedures. Part of the students' task, which is related to the research question component or criterion, is to identify the question presented in the problem. An example generation task is the design of a research project where groups of students had to produce a research question worthy of investigation. Repeated assessments of performance on these two types of tasks were collected over time.

The groups' performances on these tasks suggest that identifying and generating a research question are quite different skills. This finding confirms the assumption that producing and critiquing statistics are two rather distinct skills. Both groups had successfully acquired these skills; however, each group was stronger in one skill than the other. Metallica appeared to be more capable than Magic in identifying research questions, with the group's performance increasing steadily over time from the pretest (time 1), tutorial homework assignment 1 (time 2), and posttest (time 5), respectively (see Fig. 8.4). Metallica's performance peaked at time 2 and then slightly declined.

Magic, in contrast, was more capable at generating a clear research question and specifying all the variables along with specific levels within each variable than Metallica, outperforming the latter in all time slots except at pretesting (see Fig. 8.5). A continuous performance gain is indicated over time for Magic, with a ceiling achieved for time 4 (group project) and 5 (posttest). In general, Magic performed better over time except for low performance on the homework assignment.

Group protocols suggest that Magic's superior ability to generate an appropriate research question may be attributed to the group members' ability to integrate their own creation of a research question (i.e., favorite basketball teams) with the knowlege they acquired while viewing the library of exemplars research question criterion. An excerpt of their protocol is provided:

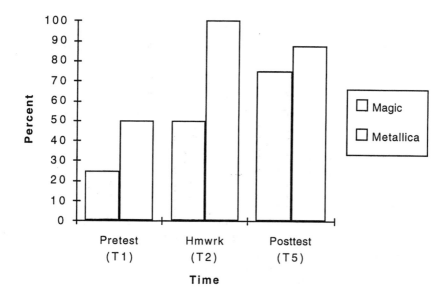

FIG. 8.4. Each group's performance on identification of a research question over five time stamps.

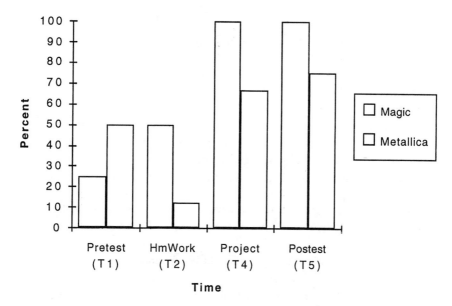

FIG. 8.5. Each group's performance on generation of a research question over five time stamps.

Magic S1:	[selects the Average criteria for Research Question] What is your favorite fast food restaurant [computer clip]? Like McDonald's
Magic S2:	Well that's what I was gonna say. What is your favorite basketball team between the Chicago Bulls, the Knicks
Tutor:	Do you see the difference between the two?
Magic S1:	Yeah, that one's really ah—the average is really ah
Magic S2:	This is—listing the question. Okay, okay, it's more like specific, it's giving you things you can choose and the average is just like—
Magic S1:	The main question.
Magic S2:	Yeah.

Magic clearly understood how to refine the research question as a result of viewing the library of exemplars, and consequently met this performance standard. Metallica also met this criterion but spent much more time selecting a question for their project, discussing seven possibilities compared to Magic's one, before settling on their first question about music preferences. The following protocol demonstrates this versatility.

MetallicaS2:	How many people like Guns and Roses?
MetallicaS1:	No that's gross . . . I don't like that.
MetallicaS3:	I don't like that.
MetallicaS4:	How many people prefer pop then. What's your favorite kind of music. Okay so how many like . . .
MetallicaS2:	Something interesting.
MetallicaS4:	How many people like . . .
MetallicaS3:	Dance music, metal, pop, rap and reggae.
MetallicaS2:	Yeah right! Dance music and metal . . . I think I like that.
MetallicaS4:	Okay let's . . . How many people prefer . . .
MetallicaS2:	What's your favorite shade of lipstick?
MetallicaS1:	No we're not going to go ask boys what their favorite shade of lipstick is.
MetallicaS2:	Well girls yeah . . . I don't know it just popped in my head.
MetallicaS1:	Okay, how many girls wear size 34B? [Giggling]
Tutor:	What are you doing?
MetallicaS4:	We are all trying to find out our question.
Tutor:	It's not that hard.
MetallicaS2:	Well, I came up with one but . . . I don't know how good it is . . . it's kind of actually stupid actually.
Tutor:	What?
MetallicaS2:	What's your favorite lipstick . . . shade of lipstick?
Tutor:	I don't think guys have any.

MetallicaS3:	Well on girls . . . on girls what they like . . .
Tutor:	or you can only ask the girls.
MetallicaS4:	Yeah! we can only ask the girls. We can ask a different question for the boys like what is your favorite basketball team?
Tutor:	Like, what you have to do is ask a question and give four or five choices
Tutor:	There where it says name, you write your question . . . name of the question like Rob he wrote types of cigarettes people prefer and he has Du Maurier, Players. That way you give choices, not choices like yes, no, or maybe.
MetallicaS4:	Oh! so we each give a question?
Tutor:	Right . . . then you choose the best one.

This versatility could be due to the fact that the students in this group have more diverse interests, or may reveal some "attention-getting" behavior or simply reflect the interests of this adolescent sample.

Data Collection. Groups were given various tasks related to data collection. For instance, paper-and-pencil tasks were somewhat constrained in that they required students to explain how they would go about collecting data to answer a particular question, such as whether or not Macintosh or IBM computers were the preferred technology by eighth-grade students in a particular province. The performance task was more flexible and required that groups of students collect data for a question they created for a statistics project. Each group's comprehension of data collection included a grasp of concepts, such as sample size, as well as how it gathered information pertaining to its question and whether or not it avoided bias in the data collection process. Assessment of each group's performance related to data collection revealed somewhat erratic patterns (see Fig. 8.6). Magic appears to be more consistent in its growth over the first four time slots, reaching 80% in time 4, but then plunging on the posttest to about 38%. A possible explanation for this sharp decline is the influence of the mode of assessment on the validity of the data. More specifically, we know from this group's protocols that they understood more about these concepts than what they presented to the class. It is possible that the posttest was another mode of assessment that did not tap into Magic's strengths. Metallica is even less consistent than Magic in terms of growth over time, but appears to do better than Magic in time 2 (tutorial) and in time 5 (posttest). Again, the mode of assessment comes into play. We know from Table 8.1 that Metallica performed better on tasks of declarative knowledge, which were emphasized on paper-and-pencil measures during times 2 and 5.

The preceding data suggest that the mode of assessment is an important consideration. However, multiple assessments taken over time are also needed. Protocols collected during the construction of statistics projects for

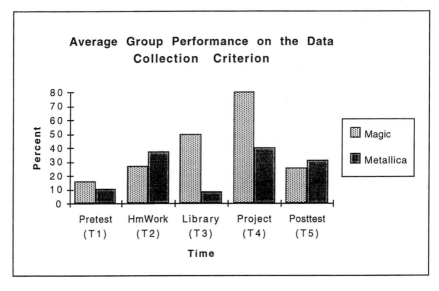

FIG. 8.6.   Each group's performance on data collection over five time stamps.

both groups revealed important information that was not presented during oral presentations in the design phase. As a group, Magic discussed three issues related to sampling: (a) age of sample—teens/eighth graders; (b) gender of sample—as many girls as boys; and (c) sampling procedure—to randomly sample eighth-grade students in their class and in other Grade 8 classes. In their oral presentation, however, the group only discussed one of these issues: age of sample. Magic, also interested in gender as a variable, failed to discuss certain key aspects of the data collection process such as the number of females and males sampled, or the total number in the sample during their group presentation. Furthermore, the sample was restricted to one age group, even though the group recognized that adults and adolescents may vary in their musical tastes.

Protocols also reveal changes in thinking as the groups engage in a performance task that might not be reflected in the creation of an artifact, in this case, a statistics project. Metallica's group discussions, for instance, revealed that the group did not initially intend to examine gender as a variable, as their presentation suggests, but decided to include this variable halfway through their design of the project. Metallica did re-collect their data and created two different graphs, one with the first set of data and one with the new data. By collecting their second data set, students demonstrated that they understood the need to collect the right kind of data to answer a particular question. The following protocol reveals the complexity that resulted from their revision of the research question in terms of the data collection process.

MetallicaS1:	We want to look at how many guys picked heavy metal . . . We want to look at no . . . , We want to know how many of each sex picked which categories.
Tutor:	Well, first you need to look at the data to see how many of each there were
MetallicaS2:	Oh, Oh! I'm not sure we have that. I know anyway just because I know these other kids, or, I know. They're all here today right now. Why don't I re-collect the data real quickly and I'll record which sex picks which category.

The benefits of multiple methods are apparent in these two cases; group discussions revealed transitions in students thinking and demonstrated that students understood more than was communicated in their oral presentations. Multiple measures can provide teachers with more evidence of student knowledge, enhancing their ability to make more informed decisions about changes to instruction. One change that we plan to make in our instruction, due to this finding, is to create a set of scripted probes that will be administered during oral presentations. Given our findings that students sometimes know more than they tell during their actual presentations, a set of standardized probes that query students on their understanding of each statistical component might give students the incentive to reveal their reasoning.

Data Analysis. Once again, paper-and-pencil and performance tasks were used that required groups to understand and calculate measures of central tendency and range and be able to determine appropriate statistics for a given problem or project, as well as interpret data and explain analyses. Figure 8.7 demonstrates that both groups had little understanding of data analysis at pretesting (time 1), but increased their knowledge during the tutorial (time 2). However, during the library of exemplars phase, Magic outperformed Metallica and continued to do so on the project (time 4) and posttest (time 5). Magic appeared to have a better understanding of data analysis procedures than Metallica. Both groups steadily increased in their performance from time 1 through time 4, but curiously decreased in their performance from time 4 to 5. Perhaps this decline reflects the groups' exclusive focus on graphs and consequent use of frequencies or percentages at the expense of concepts such as mean. Although the groups acquired some knowledge of data analysis in the tutorial, they did not fully understand the process.

Closer examination of the group journal, related specifically to statistics projects, suggests that Magic planned to compute the mean and range. However, verbal dialogues and screen recordings failed to show evidence that any of these analyses were conducted. Rather, the group focused entirely on constructing graphs to represent the data they collected. Frequencies

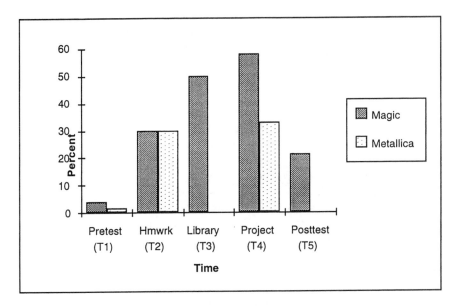

FIG. 8.7.   Each group's performance on data analysis over five time stamps.

were tabulated for each basketball team and represented in a column graph. Percentages were then automatically computed by the computer to construct a pie graph. As such, frequencies and percentages formed the basis of the group's analysis. These findings are discussed in the group's oral presentation, an excerpt of which is provided:

MagicS1:	Okay . . . um . . . for the Bulls we got 14, for the Knicks 2, Magics, 6, for the Trailblazers 0, the Celtics 4, the Suns 4, and the Spurs 0.
MagicS2:	And that was the number of people out of 30.
MagicS2:	Okay . . . we also did a pie graph to show um . . . the percentages but um . . . we feel the other graph is easier to understand because the people out of thirty show it better . . . what we found better. Any ways, the Bulls are 46.67%; the Knicks are 6.67%; the Spurs are . . . the Magics are 20%; the Celtics 13.33%, and the Suns are 13.33%.
Tutor:	So um . . . who was . . . who won . . . who was the least favorite and who was the most favorite?
MagicS2:	The least favorite . . . well um . . . like a . . . since pie graphs can't show zero values umm . . . like a . . . Trailblazers and Spurs are not in the pie chart.
Tutor:	Did you divide your data . . . you said you collected your data from three different classrooms.

MagicS1:   This class and other classes.

Tutor: But did you look at those classes versus this class? Do you
             think it would make a difference when you asked . . . whether
             or not popularity would change?

MagicS2:   No . . . umm

MagicS1:   Some people might have different opinions . . . they might like
             another team better.

Notice that the results were presented in the form of numbers (frequencies
and percentages), but were not qualified in terms of "least favorite" and
"most favorite" terminology that reflect the group's research question, that
is, what is your favorite basketball team. In this sense, the group did not
adequately explain its results until questioned further.

Metallica was able to describe the group's results, but not extrapolate to
explain what the findings meant in the context of their population. The
group discussed the magnitude of differences between music selections but
never in terms of percentages or frequencies, simply most or least. Limited
use of statistics by both groups was also reflected in the screen recordings,
which indicated a low frequency of statistical manipulation over time. Figure
8.8 illustrates the number of instances where Magic and Metallica entered
data, transformed data, and produced statistics from the first 3 days of the
tutorial (T1, T2, T3) and the first day of the project (T4). As with the paper
and pencil and video assessments, the two groups diverged over time, with
Magic performing more statistical manipulations than Metallica. However,
the number of statistical manipulations decreased for both groups during
the projects, with most of the manipulations being related to entering data.

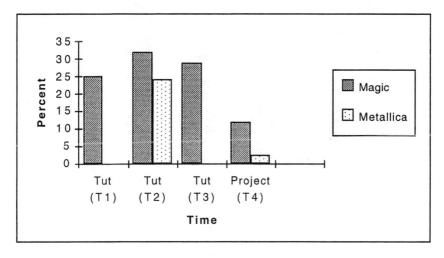

FIG. 8.8. Each group's performance on statistical manipulations over time.

It is clear from the assessments and the screen recordings that neither group explored data analysis procedures in any depth—hence their decreased performance on the posttest.

Data Representation. Data representation refers to how data are summarized, presented, and interpreted and whether or not the types of tables, charts, and/or graphs that students constructed to represent data were appropriate. One of the key features of the ASP instruction was the use of graphs for representing variables and data graphically. The tutorial provided instruction on how to use computer software (i.e., CricketGraph) for representing and interpreting data. The class was familiar with some of the graphs (i.e., column graph, bar graph, pie chart, and line graph) taught in ASP but not others (i.e., area graph, scatterplot, and area graph). Students were expected to use their knowledge of graph construction and their skill in interpreting graphs in the pre- and posttest, homework assignment 1, and for representing the data collected for their research project. Figure 8.9 reveals that Metallica started out lower in these skills than Magic but steadily increased its performance over time and finished at a higher level of performance than Magic at time 5 (posttest). Magic performed well but dropped slightly from time 4 (project) to time 5 (posttest).

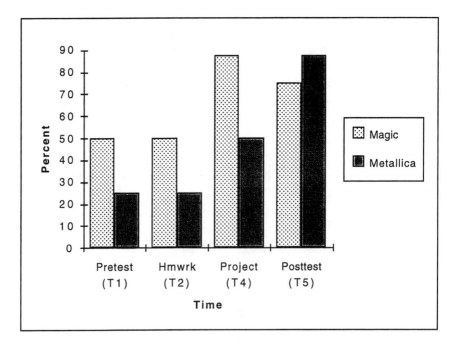

FIG. 8.9. Each group's performance on data representation.

We also examined the groups' screen recordings for number and type of graphical manipulations over time and found that the frequency with which the groups manipulated graphics in the tutorial was low but increased during the project phase of ASP (see Fig. 8.10). This finding is not surprising given that the first three tutorial sessions focused on entering and analyzing data. Graphics were not introduced until the fourth tutorial session. Nonetheless, the increased frequencies at time 4 suggest that both groups had an interest in producing and modifying graphics for their statistics projects, which may have resulted from this last tutorial session. This interest in graphics for both groups was also illustrated in the oral presentations.

Magic and Metallica spent a great deal of time personalizing their graphics by improving such details as fonts, patterns, and backgrounds. They took ownership of their work by labeling or saving graphical files based on their group names. The screen recording data indicate that Metallica spent considerably more time modifying the graphics than Magic. The screen recording data also provided valuable information regarding off-task actions and student difficulties. Of the total actions engaged in by Metallica, 10% were off task. Closer examination of the group's dialogues and its computer actions suggests that this off-task behavior may in part be attributed to difficulties in using the computer software.

Verbal protocols and the screen recordings indicate that Magic tried three types of graphs: a line graph, a column graph (not a bar graph), and a pie chart. The group decided to keep and modify the column and pie graphs. Again, the protocols and screen recording provide a comprehensive profile.

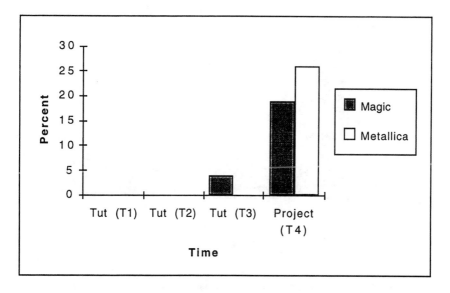

FIG. 8.10. Each group's performance on graphical manipulations over time.

The screen recording provides a visual representation of the graph in various stages of design. The changes that students make are of interest. The group's identification of the axes, for instance, demonstrates their understanding of graphing techniques as well as of what the representation itself means. The *y*-axis label was appropriately changed from tally to "# of people" because the data are based on frequencies (see Fig. 8.11). The group chose to entitle the graph "favorite basketball teams." In a sense, the graph is adequately labeled. Usually titles for graphs identify the independent and dependent variables. These are identified by "basketball teams" and "favorite," respectively. The title, however, lacks information about which group of people the data represents.

Magic also created a pie chart (see Fig. 8.12) illustrating the percentage of eighth-grade students that selected a particular basketball team as their preferred team. The group encountered difficulties due to the zero values obtained for two of their basketball teams (i.e., Trailblazers and Spurs), which could not be represented in the pie chart. The group consequently deleted these values from the datafile and was able to successfully create its pie chart using the remaining five basketball teams.

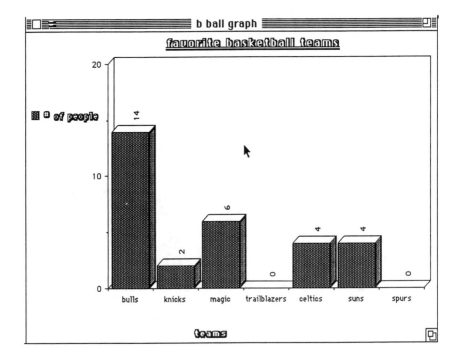

FIG. 8.11. Magic's construction of a column graph that represents its project data.

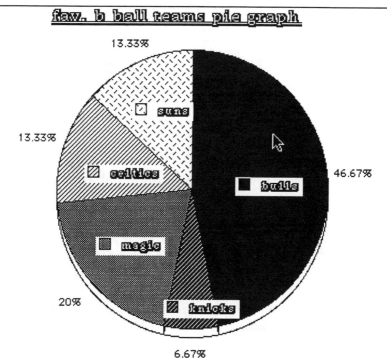

FIG. 8.12. Magic's pie graph that represents its project data.

The group presented its project using the column and pie graphs created to represent the data. These representations served as the basis for interpreting the data. The following dialogues demonstrated the group's description and explanation of the superiority of the column graph (see Fig. 8.11) over the pie graph (see Fig. 8.12).

MagicS2: Okay . . . we also did a pie graph to show um . . . the percentages but um. . . we feel the other graph [column] is easier to understand because the people out of thirty show it better . . . what we found better. Anyways, the Bulls are 46.67%; the Knicks are 6.67%; the Spurs are . . . the Magics are 20%; the Celtics 13.33%, and the Suns are 13.33%.

Metallica created two graphs, one for overall music preferences (see Fig. 8.13) and one for preferences by gender (see Fig. 8.14). The first graph was a pie graph and the second a column graph. The group isolated "dance" as preferable in the pie graph and then described the different preferences by gender in the column graph.

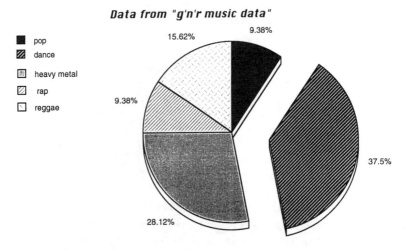

FIG. 8.13. Metallica's pie chart on overall preferences that represents its project data.

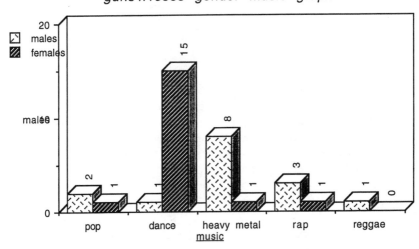

FIG. 8.14. Metallica's column graph of preferences by gender that represents its project data.

DISCUSSION AND FUTURE DIRECTIONS

We have demonstrated ways in which student progress in statistics can be monitored in ASP. The theoretical basis of ASP was discussed, and our premise was that authentic, project-based learning environments, where students "do" statistics that are meaningful to them, can facilitate statistical understanding. A culture of explaining evolved in the ASP classroom, and our multiple methods of assessment allowed us to document changes in students' statistical knowledge, in terms of declarative, procedural, and conceptual knowledge.

An in-depth examination of two cases was presented to emphasize the utility of multiple forms of assessment for monitoring student progress. The groups were similar in some ways and different in others. Magic demonstrated more conceptual knowledge in statistics than Metallica, but Metallica was stronger in declarative knowledge than Magic. Both groups had a similar profile for procedural knowledge. Our intent in presenting this in-depth analysis is to demonstrate ways in which researchers and teachers can build theories or hypotheses as to how students learn statistics. Once we understand the process of learning, it may be easier to formalize better forms of instruction and assessment in statistics. For example, by identifying how individuals differ in terms of type of knowledge, instruction can be designed to make instructional decisions based on an individual's or group's strengths and weaknesses in a particular area of statistics. Our in-depth analysis of Magic and Metallica demonstrated such strengths and weaknesses and helped us identify which types of assessment measures were best for whom. Although there is a great deal of complexity in such data, we believe that in time, we will be able to create some recommendations for teachers that will make the assessment tasks more manageable, and understanding learners more apparent.

A major goal of assessment is for teachers to monitor student progress so that they can make important instructional decisions that will improve individual learning (NCTM, 1995). It is just as important that students learn to monitor their own progress and become independent assessors of their own knowledge. Teachers can monitor student progress on the assessment criteria and examine whether or not students' self-assessments are in accord with their own assessments of student work. Such an exercise would verify that teachers are communicating their goals clearly to students and would also provide a checking mechanism for both the teacher and student that learning is occurring. We have implemented three types of assessments that are designed to assist both teachers and students to learn from the assessment experience. Teachers learn what their students do or do not understand. Consequently, teachers can alter their instructional strategies accordingly. Students learn what they know or do not know and can seek assistance when they need more information to understand something more completely.

In ASP, teachers assessed groups on each statistical criterion. Second, members of each student group rated themselves on the criteria by discussing their performance and reaching a consensus in their assessment. Finally, each group assessed other group projects in the same manner. Our work with high-school students suggests that making criteria open and visible to groups through technology helps them to closely align their assessment of their own work with the teachers' assessment. One way to facilitate this independence is to make the instructional and assessment goals clear to the learner. In the ASP, this was conducted via a series of exemplars of "average" and "above average" performance. These were provided as performance standards for the components of statistical problem solving. It is our premise that when students are made aware of what is expected of them, it easier for them to meet expectations, to assess their own progress, and to compare their work with others. Students can also learn about their strengths and weaknesses, as well as gaps in knowledge. Such information can provide a way for students to take charge of their own learning.

This chapter has provided statistical tasks that are authentic to the student, teacher, and assessor. They were authentic to the (a) student because the research questions were generated by the students themselves, and thus they took ownership of solving a statistical problem; (b) teacher because the activities embodied the NCTM guidelines for statistics, that is, providing opportunities to reflect, organize, model, represent, and argue within and across mathematical domains; and (c) assessor because they provided multiple indicators of what students understood and provided ongoing assessment opportunities that allowed the assessor to monitor students' growth over time.

It is evident that performance standards, when established and made open to a student prior to performance, can assist students in achieving the instructional goals set forth in the standards. When students are made aware of what is expected of them, it is easier for them to meet those expectations. Technology, as demonstrated in the library of exemplars, can be used to make such expectations clear. Technology can help make abstract criteria more concrete and provide benchmarks or anchors for statistical problem solving (see Lavigne, 1994, for a full description of the study). Students can use such criteria to assess their own progress and compare their work with others. Consistent evidence from all of our "assessors" enhances the validity of our inference that students using ASP understood the statistical performance standards and how to apply their understanding to their own projects.

As shown in Fig. 8.1, many types of assessments were used and different scoring procedures were developed. Scoring and scaling procedures were constructed in ways appropriate to the assessment tasks. From reviewing the case profiles, it is obvious that different types of assessments are more conducive to some students than for others and that without these multiple sources of evidence valid measures of the depth and breadth of student knowledge would not have been documented. This research is only a first step at identifying the appropriate assessment tasks for learning descriptive statistics with ASP. Our future studies will concentrate on improving these assessment measures and making the management and scoring of these tasks more teacher-friendly.

We also found support for our assumption that allowing student input regarding research questions would make statistics a more meaningful experience. Students liked using computers to learn statistics and enjoyed working in groups and conducting their own studies. As Blumenfeld et al. (1991) suggested, cognitive engagement does increase when students are responsible for the creation of their own questions and activities. Our evaluations indicate that such a relationship exists, but more empirical study of this relationship is required, particularly in terms of maintaining motivation and learning over time. Perhaps there is a link that can be further studied between students' intrinsic motivation and learning statistics.

ASP demonstrated that multiple sources of evidence of learning can be collected through different modes of assessment and that such measures allow for more valid indications of what was learned. In a sense, these multiple means of assessment were adaptive to individual differences in learning in that certain assessment modes were better for some groups than for others. These results support the assumption that there is no best form of instruction or assessment for all individuals (Snow, 1989). Armed with these new types of assessments, teachers have more evidence of student knowledge and can make more informed decisions about changes to instruction. Just as importantly, students can make more informed changes to their learning strategies when assessment goals are clear to them.

ACKNOWLEDGMENTS

Preparation of this chapter was made possible through funding from the Office of Educational Research and Improvement, National Center for Research in Mathematical Sciences Education (NCRMSE). Special thanks to Dr. Thomas Romberg for his support and encouragement of this research endeavor. Additional thanks to Phil Knox, a gifted teacher, who made our data collection possible. Special thanks to André Renaud for his programming skills. Further acknowledgments to Marlene Desjardins, Cindy Finn, Bryn Holmes, Jody Markow, Tina Newman, Litsa Papathanasopoulou, and Steve Reisler for volunteering their time throughout the study.

REFERENCES

American Statistical Association. (1991). *Guidelines for the teaching of statistics K–12 mathematics curriculum.* Landover, MD: Corporate Press.

Anderson, J. R. (1983). *Cognitive psychology and its implications.* New York: Academic Press.

Berliner, D. C. (1992). Telling the stories of educational psychology. *Educational Psychologist, 27*(2), 143–161.

Blumenfeld, P. C., Soloway, E., Marx, R. W., Krajcik, J. S., Guzdial, M., & Palincsar, A. (1991). Motivating project-based learning: Sustaining the doing, supporting the learning. *Educational Psychologist, 26*(3&4), 369–398.

Blumenfeld, P. C., Krajcik, J. S., Marx, R. W., & Soloway, E. (1994). Lessons learned: How collaboration helped middle grade science teachers learn project-based instruction. *Elementary School Journal, 94*(5), 539–551.

Bransford, J., Hasselbring, T., Barron, B., Kulewicz, S., Littlefield, J., & Goin, L. (1988). Uses of macro-contexts to facilitate mathematical thinking. In R. Charles & E. A. Silver (Eds.), *The teaching and assessing of mathematical problem solving* (pp. 125–147). Reston, VA: National Council of Teachers of Mathematics.

Carter, K. (1993). The place of story in the study of teaching and teacher education. *Educational Researcher, 22*(1), 5–12.

Cognition and Technology Group at Vanderbilt. (1992). The Jasper series as an example of anchored instruction: Theory, program description, and assessment data. *Educational Psychologist, 27*(3), 291–315.

Collins, A., Brown. J. S., & Newman, S. (1989). Cognitive apprenticeship: Teaching the craft of reading, writing, and mathematics. In L. B. Resnick (Ed.), *Cognition and instruction: Issues and agendas* (pp. 453–494). Hillsdale, NJ: Lawrence Erlbaum Associates.

DuCharme, C. C. (1993, November). *Historical roots of the project approach in the United States: 1850–1930.* Paper presented at the Annual Convention of the National Association for the Education of Young Children, Anaheim, CA.

Edgerton, R. T. (1993). Apply the curriculum standards with project questions. *Mathematics Teacher, 86*(8), 686–689.

Frederiksen, J. R., & Collins, A. (1989). A systems approach to educational testing. *Educational Researcher, 18*(9), 27–32.

Gorb, P. (1987). Projects not cases: Teaching design to managers. *Management Education and Development, 18*(4), 299–307.

Kant, I. (1968). *Critique of pure reason* (Norman Kemp Smith, Trans.). New York: St. Martin's Press. (Original work published 1787)

Krajcik, J. S., Blumenfeld, P. C., Marx, R. W., & Soloway, E. (1994). A collaborative model for helping middle grade science teachers learn project-based instruction. *Elementary School Journal, 94*(5), 483–497.

Ladewski, B. G., Krajcik, J. S., & Harvey, C. L. (1994). A middle grade science teacher's emerging understanding of project-based instruction. *Elementary School Journal, 94*(5), 499–515.

Lajoie, S. P. (1996). The use of technology for modeling performance standards in statistics. *Proceedings of the International Association for Statistical Education Roundtable on Research on the Role of Technology in Teaching and Learning Statistics* (pp. 69–83). Granada, Spain.

Lajoie, S. P., Jacobs, V. J., & Lavigne, N. C. (1995). Empowering children in the use of statistics. *Journal of Mathematical Behavior, 14*(4), 401–425.

Lavigne, N. C. (1994). *Authentic assessment: A library of exemplars for enhancing statistics perfromance.* Unpublished masters thesis, McGill University, Montreal.

Lavigne, N. C., & Lajoie, S. P. (1996). Communicating performance standards to students through technology. *Mathematics Teacher, 89*(1), 66–69.

National Council of Teachers of Mathematics. (1989). *Curriculum and evaluation standards for school mathematics.* Reston, VA: Author.

National Council of Teachers of Mathematics. (1995). *Assessment standards.* Reston, VA: Author.

Nitko, A. J. (1989). Designing tests that are integrated with instruction. In R. L. Linn (Ed.), *Educational measurement* (3rd ed., pp. 447–474). Washington, DC: Macmillan.

Resnick, L. B. (1987). Learning in school and out. *Educational Researcher, 16*, 13–20.

Romberg, T. A. (1992). Evaluation: A coat of many colors. In T. A. Romberg (Ed.), *Mathematics assessment and evaluation: Imperatives for mathematics educators* (pp. 10–36). Albany, NY: SUNY Press.

Schoenfeld, A. (1985). *Mathematical problem solving.* New York: Academic Press.

Scott, C. A. (1994). Project-based science: Reflections of a middle school teacher. *Elementary School Journal, 95*(1), 75–94.

Shute, V. J., & Gawlick-Grendell, L. A. (1994). What does the computer contribute to learning? *Computers and Education, 23(3)*, 177–186.

Snow, R. E. (1989). Toward assessment of cognitive and conative structures in learning. *Educational Researcher, 18*(9), 8–14.

Sommers, J. (1992). Statistics in the classroom: Written projects portraying real-world situations. *Mathematics Teacher, 85*(4), 310–313.

Trepanier-Street, M. (1993). What's so new about the project approach? *Childhood Education, 70*(1), 25–28.

von Glasersfeld, E. (1988). An exposition of constructivism: Why some like it radical? *Journal of Research on Mathematics Education Monograph, 4,* 19–30.

Vygotsky, L. S. (1896–1934/1978). *Mind in society: The development of higher psychological processes* (M. Cole, V. John-Steiner, S. Scribner, & E. Souberman, Eds. and Trans.). Cambridge, MA: Harvard University Press.

Wiggins, G. (1992). Creating tests worth taking. *Educational Leadership, 49*(8), 26–33.

Wolk, S. (1994). Project-based learning: Pursuits with a purpose. *Educational Leadership, 52*(3), 42–45.

CHAPTER NINE

Aligning Everyday and Mathematical Reasoning: The Case of Sampling Assumptions

Daniel L. Schwartz
Susan R. Goldman
Nancy J. Vye
Brigid J. Barron
The Cognition and Technology Group at Vanderbilt
*Learning Technology Center, Vanderbilt University,
Nashville, Tennessee*

In this chapter we present results from three studies that examined and supported fifth- and sixth-grade children's evolving notions of sampling and statistical inference. Our primary finding has been that the context of a statistical problem exerts a profound influence on children's assumptions about the purpose and validity of a sample. A random sample in the context of drawing marbles, for example, is considered acceptable, whereas a random sample in the context of an opinion survey is not. In our design of instructional and assessment materials, we have tried to acknowledge and take advantage of the role that context plays in statistical understanding.

The chapter has three main sections. In the first section, we present evidence on the piecemeal and context-sensitive nature of statistical understanding. We use this as a basis for proposing that statistical instruction should often be situated in everyday contexts. In the second section, we examine the commonsense basis of early statistical understanding. We present evidence that children understand statistical contexts by drawing on schemas for more familiar situations, such as advertising, that have family resemblances to various aspects of a statistical inference. In the third section, we describe the results of our efforts to build on children's piecemeal, context-sensitive common sense. Our approach has depended on a complex video-based problem from *The Adventures of Jasper Woodbury* (Cognition and Technology Group at Vanderbilt [CTGV], 1992a, 1992b) that situates statistical instruction in an everyday context (CTGV, 1990; Moore et al.,

1994). Folded into this instruction is an assessment model that helps children make their assumptions visible and amenable to revision (Barron et al., 1995). The results show that children's multiple everyday understandings about statistical inference can be improved by eliciting and anchoring their diverse intuitions within the same context.

NAIVE STATISTICAL REASONING FROM AN INSTRUCTIONAL PERSPECTIVE

The Piecemeal Character of Common Sense

We assume that mature statistical understanding evolves from everyday intuitions—intuitions that are strongly affected by the context in which problems and examples are situated. We believe that commonsense understanding is often best characterized as pockets of disconnected intuition with piecemeal application (Heider, 1958). It is only after considerable effort that these pockets develop the consistency of theoretical knowledge. This assumption about how to characterize everyday understanding is different from the assumption that understanding should be described as a set of coherent rules or principles that characterize one's theory of some overall domain. diSessa's (1983, 1993) construct of "knowledge in pieces" in the domain of physics provides an excellent example of the position we wish to emphasize. He argued that people's understanding of the physical world comes in discrete pieces of intuitive understanding whose elicitation is contingent upon the problem context. "Scientific explanation begins with common sense observation, a principal characteristic of which is its appearance as disparate and isolated special cases" (diSessa, 1983, p. 16). Although experts may have well-developed, coherent sets of principles, novices do not (Chi, Feltovich, & Glaser, 1981; Larkin, McDermott, Simon, & Simon, 1980). Under this model, conceptual growth does not begin with first principles, such as the laws of thermodynamics, that are subsequently mapped into specific cases. Rather, the growth of understanding is characterized as a process of sifting through and reconciling the cases, "finding successively the more and more general and fundamental ones which serve as principles, explaining the more special cases" (diSessa, 1983, p. 16). In this chapter, we suggest that learning and understanding in statistical domains should also be characterized as the integration of pieces of contextual knowledge into fuller understanding.

Our approach to examining the commonsense basis of statistical understanding contrasts with a dominant theoretical approach in which statistical understanding is compared to the formal knowledge of the expert. In this approach, embodied in the judgment under uncertainty literature (see

Kahneman, Slovic, & Tversky, 1982), the emphasis is on people's misconceptions about the normative principles of statistical inference. This approach has led to great insight on the limitations of people's ability to reason statistically. It does not, however, readily yield an understanding of the development of normative statistical reasoning nor prescriptions for instructional practice. In part, this is because the literature has endeavored to show that people fail to reason normatively. Consequently, the research has been relatively silent about the contexts and commonsense understandings that may support the development of normative statistical reasoning.

Tversky and Kahneman (1983), for example, provided subjects with a profile of 31-year-old Linda who was described as single, outspoken, bright, philosophical, and deeply concerned about issues of social equity. The subjects were asked which of two alternatives were more probable: (a) Linda is a bank teller, or (b) Linda is a bank teller and is active in the feminist movement. Most subjects chose option (b), although option (a) is more likely in a statistical sense. The error, statistically speaking, is a version of the conjunction fallacy in which conjoint events are considered more likely than one of the events alone (i.e., a feminist and banker vs. a banker). Even though the question asks for the most *probable* outcome, subjects tended to interpret the problem in nonstatistical terms. They chose the feminist and banker option because it had the greatest overlap, or covariation, with Linda's description. It seemed more representative (Kahneman & Tversky, 1972) or prototypical of someone like Linda. By norming people's performance on the Linda question against the coherent knowledge of the statistical expert, their reasoning appears as a misconception; people do not understand conjunctive probabilities. From our perspective, however, people are not exhibiting a misconception so much as a misapplication. They are revealing how commonsense understanding is dependent on specific contexts of application (Donaldson, 1978; Johnson-Laird, Legrenzi, & Legrenzi, 1972). People's errors on the Linda problem do not necessarily imply that they do not understand conjunctive probabilities. Given a problem that asks which is a more probable event, (a) a male or (b) a male with black hair, we suspect that most older children and adults would correctly choose option (a). Thus, people do not have misconceptions about conjunctive probabilities, they simply have competing understandings. Knowledge relevant to answering a question about a "probable" event is differentially applied depending on the pocket of understanding that is drawn on for a particular situation.

Ideally, people would not switch from one fragment of knowledge to another depending on incidental problem features. Instead, they would have a more comprehensive body of knowledge that supports judicious interpretations. Coherent understanding, however, is the endpoint of instruction, not the starting point. We believe statistical instruction should directly grapple with the context-sensitive and piecemeal nature of common sense and use it to advantage. Rather than using instruction that begins with abstract principles (e.g., conjunctive probabilities) and disregards the contexts that give rise to competing intuitions, instruction should include situations in which competing intuitions can emerge and be reconciled into a more coherent body of understanding. For example, instead of using an abstract lesson on conjunctive probability with a concrete example or two, one might begin with both the Linda and the male-with-black-hair scenarios. One might ask students to differentiate and reconcile the reasons they give different answers to each problem, and then introduce the abstract principle of a conjunctive probability as it applies to both situations. In sum, the world should not serve as an example for applying an abstract mathematical principle. Rather an abstract principle should serve as an example for how to model the world.

Situating Common Sense in Instruction

The idea that different contexts can elicit competing intuitions suggests a novel reason for contextualizing instruction in everyday situations. To develop this point, it is worth first considering previous arguments for situating mathematics instruction. To our knowledge, the arguments for contextualizing mathematics instruction have pointed to properties of situations that naturally scaffold desirable mathematical thinking and practices. There have been four broad lines of theorizing and evidence. The *culture of mathematics* position states that learning should take place in appropriate social contexts, because learning involves the development of mathematical social norms (e.g., Cobb, Wood, Yackel, & McNeal, 1992; Newman, Griffin, & Cole, 1989) and access to models of conventionalized practice (e.g., Brown, Collins, & Duguid, 1989; Lave & Wenger, 1991). The *inert knowledge* position (cf. Whitehead, 1929) argues that people should learn mathematics in context so they can later recognize situations that are susceptible to the tools of mathematics (e.g., Bransford, Franks, Vye & Sherwood, 1989; Bransford, Sherwood, Hasselbring, Kinzer, & Williams, 1990). The *expertise* position claims that rich contexts support multiple revistings to complex ideas. This yields expert-like flexibility, planfulness, and insight into deep structure not available with abstracted and preframed worksheet problems (e.g., Spiro, Feltovich, Jacobson, & Coulson, 1991). Finally, the *affordances* position argues that instruction can capitalize on people's understandings of and interactions with the structure of the physical world to help structure their mathematical understanding (e.g., Greeno, Smith, & Moore, 1993). At their core, each of these four positions depends on the structure of the child's experience and activity to provide a natural pathway into mathematical competence.

Although we agree with the thrust of these positions, we add a fifth, contrasting reason for contextualizing instruction that we label the *incongruency* position. In some situations, commonsense understandings of the everyday world do not easily map onto mathematical understandings (e.g., Konold, 1989). In these situations, the connection between everyday and mathematical reasoning needs to be developed through the tandem processes of discrimination and reconciliation, allowing both bodies of knowledge to inform one another.

Turning Situations Into Mathematics:
The Role of Measurement

The case in point for the incongruency position involves statistical inference. Statistical inference is a challenging domain for developing the connection between mathematical and everyday understanding. As previously noted, adults do not apply normative statistical reasoning in many situations (Kahneman, Slovic, & Tversky, 1982). One possible explanation is that statistics has a second layer of complexity compared to many other forms of mathematical application. In general, the connection between the experienced world and mathematics depends on a system of measurement that converts phenomena into numbers. For example, one can turn the sides of a square into a number of inches. Afterward, one can multiply the numerical values to model the physical area of the square. For mathematics to model situations accurately, measurement systems need to satisfy assumptions that allow them to serve as theoretically sound bridges between the phenomenal world and mathematics. Many systems of mathematics, for example, require that measurements show transitive relationships such that if measurement A is greater than measurement B, and B is greater than C, one will always find that A is greater than C. Otherwise, certain mathematical operations cannot be guaranteed to model the world accurately. To apply statistics, there are also systems of measurement that serve as the bridge between mathematics and phenomena. In particular, a sample provides the measurement of the distribution needed for a statistical analysis. Like all systems of measurement, samples need to satisfy a number of assumptions to fit within particular mathematical models (e.g., normality and independence for a parametric analysis). Sampling, however, is psychologically unlike other systems of measurement for two reasons.

The first reason is that assumptions about the properties of a sample often refer to the distribution of cases that comprise the sample. The assumptions occur at the level of aggregates rather than more easily imagined and observed single instances. For example, although it is not too difficult to imagine a prototypical instance of a lazy person, it is fairly difficult to imagine a distribution of lazy people, short of thinking in terms of a histogram. We suspect that college instruction does not spend enough time developing the assumptions that must be met for a mathematical treatment of aggregates. We can recount multiple instances, at multiple universities, when advanced students have not actively recognized that the usefulness of the mean or standard deviation depends on specific distributional assumptions.

The second reason that sampling is psychologically unlike other forms of measurement—the reason that we focus on in this chapter—is that sampling is a measurement of measurements. For example, in a psychology experiment, subjects produce a behavior that gets measured (e.g., a reaction time), and the researcher measures the distribution of the target behavior (e.g. a sample of reaction times). This dual measurement means that one must navigate through the complexity that subjects cause the outcomes of interest (e.g., reaction time), but the way the sample of outcomes is selected must not, in and of itself, produce the distribution of outcomes; that is, the sample must be random. This juxtaposition of randomness and causality in a sampling situation lends itself to an incongruence between everyday and mathematical interpretations of a situation. People may impose the causal interpretations of everyday reasoning onto mathematical worlds in which chance is the operative rule. The following section develops this idea in detail.

Cause and Chance in Everyday Reasoning

We believe that people have difficulty separating out the causal and random elements within a statistical inference. Our point here is not only that people view chance factors as annoyances that prevent perfect prediction (Kuzmak & Gelman, 1986), or that they learn of chance as the antithesis of causality (Owens, 1992; Piaget & Inhelder, 1975). In addition, people may have trouble assigning complementary roles to the causal and random elements within a statistical inference, as would be necessary when using a random sample to test for causal relations. For example, in one of the video-based problems we have used in the classroom, the story protagonists take surveys at three schools to determine if there is sufficient support for an environmental project. At each school, they select 80 children using a different randomizing device: a spinner, a die, and names drawn from a hat. On hearing of this, a colleague suggested that we had erred because the three random samples were not generated in the same way. Our colleague had imported an inter-esting interpretation by which different "random" devices would "cause" different distributions of opinions. Evidently, even the use of different ran-domizing devices can lead to a confusion between the roles of chance and cause in producing a distribution of outcomes.

One manifestation of the confusion between cause and chance appears when people overlook the role of randomness when trying to identify causal relationships. A study by Nisbett, Krantz, Jepson, and Kunda (1983) nicely exemplifies this phenomenon. Adult subjects read about a young man who was deciding whether to go to a small liberal arts college or a large Ivy League university. The subjects' task was to make a recommendation for the prospective student based on the evidence available to the young man. In one condition, the subjects read how the young man had received information from his older friends at the two colleges. The friends at the liberal arts school reported that they liked it very much and found it very stimulating. The friends at the Ivy League school reported that they had many complaints on both personal and social grounds. However, when the young man visited the liberal arts college, he met several cold and unpleasant students and an abrupt and disinterested professor. At the Ivy League school, several people were enthusiastic and pleasant, and two professors took a personal interest. Under this version of the story, 74% of the subjects recommended that the student should go to the Ivy League university. Presumably, the subjects reasoned that the young man's experience at each university was caused by enduring characteristics of the respective institutions. In a second experimental condition, the story was changed slightly to help subjects attend to the chance factors of the young man's visits. The story described how the young man made a list of all the different places and activities he might visit, and how he dropped a pencil on his list to pick out a subset of the possibilities. Otherwise, the subjects in this condition received an identical story. The number of subjects who recommended that the young man go to the Ivy League school dropped to 56%. In this condition, as compared to the first, subjects were more likely to consider properties of the measurement selection. When chance elements were made explicit, subjects applied the law of large numbers to make their decision. They reasoned that the friends would have a much larger sample from which to measure the qualities of the two schools compared to the single visits of the young man, single visits that would be prone to the error inherent to a small sample (e.g., he just had an unlucky visit). When the chance elements of the young man's visit were implicit, subjects were inclined to disregard them and to reason that the bad visit at the small college was caused by consistent problems with the small college.

For adolescents who do not have a strong grasp of random distributions, the tendency to import causal attributions may be even greater than that observed by Nisbett (Nisbett et al., 1983; cf. Fischbein, 1975). Given the centrality of sampling as the bridge between mathematics and the world, clarity about the roles of chance and cause in sampling is an important element of statistical understanding. Nonetheless, there has been minimal developmental research that has examined how probabilistic and causal understandings interact in everyday statistical inferences (Garfield & Ahlgren, 1988; Shaughnessy, 1992). Most research into children's knowledge of chance has used materials in which the sampling procedure has a limited causal interpretation (e.g., Dean, 1987, Yost, Siegel, & Andrews, 1962). Activities like rolling dice, twirling a spinner, or drawing marbles from an urn offer limited opportunities for a causal explanation of the outcomes. It would, for example, be highly unlikely for someone to reason that a red marble *causes* its redness. These chance setups have yielded important information about children's probabilistic reasoning (e.g., Gal, Rothschild, & Wagner, 1990; Rubin, Bruce, & Tenny, 1990). This information, however, does not explain how children's understanding of sampling evolves to handle more contextual statistical situations (Jacobs & Lajoie, 1994; Nisbett et al., 1983). The move from understanding the relationship between a sample and a population in games of chance may not naturally lead to an understanding of more everyday sampling situations like taking a survey. For example, unlike the case of drawing marbles from an urn, the opinions people hold are often attributed to them *because* of some personal characteristic such as age. Consequently, children may be inclined to think about sampling characteristics of the population with the idea that those characteristics effect the distribution of opinions. As we develop in the next section, this is quite different from thinking about sampling the distribution of opinions in the population, regardless of who holds those opinions.

Thinking of Causes in Situations of Chance: The Covariance Assumption

In everyday situations, people have a tendency to overattribute causality (Kelley, 1973). People often assume causal relations solely on the basis of co-occurrence or temporal association (Goldman, 1985). It is as though people bring an assumption that events should be explained causally, and they search for covariances between events and/or properties that can support this form of explanation. We call this focus on causal explanation the *covariance assumption.* In this section, we develop evidence that the covariance assumption leads children to think of sampling in the context of a survey in terms of causing a distribution, even though they can think of sampling in random terms in contexts that do not lend themselves to causal attribution. The idea here is not that the children necessarily think of a survey as a way to find covariances, although they often do. Rather, the availability of covariances in a situation has the side effect of leading children to think of sampling in terms that cause (bias) a distribution.

To illuminate how the covariance assumption may lead to an incongruency between probabilistic and everyday reasoning, we decompose the structure of a statistical inference. Consider the three representations of statistical inference shown in Fig. 9.1. The top panel represents a typical chance setup such as pulling marbles from an urn. The left circle represents the population of marbles from which the sample is drawn. The heavy arrow in the left of the panel represents the sampling procedure—in this case, reaching into the urn and blindly pulling marbles. The rounded box represents the resulting sample. The dashed arrow represents the inference from the sample back to an estimate of the population. The inferred population distribution is represented in the circle on the right.

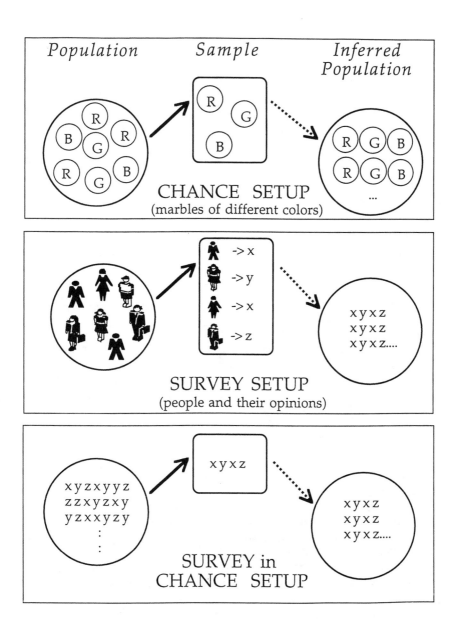

FIG. 9.1. Why a survey is psychologically unlike drawing marbles from an urn. In a survey, there is a potential covariant relationship between people's characteristics and their opinions (x, y, and z). In a chance setup, such as pulling colored marbles from an urn, one would not normally say that a red marble covaries with a red marble. From Schwartz & Goldman (1996). Reprinted by permission from John Wiley & Sons, Ltd., New York.

Next consider a statistical inference in the context of a survey. The middle panel captures a commonsense interpretation of a survey. In the left circle, there is a population of people. As before, the heavy arrow indicates the selection of the sample. In this case, however, one samples from a population of *people* who then generate *opinions*. This is represented in the center box with the small arrows indicating a relationship between the chosen individuals and their opinions, x, y, and z. The sample of opinions may then be used to infer the distribution of opinions within the original population. When viewing a survey this way, one takes two measurements: a sample of population characteristics, and a measurement of how individuals in the sample respond to the survey questions. The bottom panel shows a less intuitive way to view a survey. Here, similar to the chance setup, one samples a population of opinions to make an inference about a population of opinions.

Consider the psychological difference between the chance and survey setups. In the chance setup, the relationship between the sampled population, the sample, and the inferred population is one of self same: red marble → red marble → red marble. There is no explanation for the fact that a red marble "leads to" a red marble. In contrast, in the survey setup, there are two distinct entities: people and opinions. The categorical difference between people and opinions may affect one's interpretation of sampling. One can easily question whether something about a person "leads to" opinion x. More generally, one might view the people in the middle panel as reflecting identifiable traits within the population (e.g., age, gender, political affiliation). Consequently, one might question whether a particular trait is associated with opinion x. Or, one might question whether all the classes of people who have different opinions are represented in the sample (e.g., the figure with the baseball cap is left out). We hypothesize that the way people represent an opinion survey (as composed of people and opinions) leads to a covariance assumption. The covariance assumption invites people to focus on the traits of the population members. This focus sharpens a concern over the inclusion of those "traits" in any sample of the population. For a sophisticated statistical reasoner, this concern may be satisfied through a stratified-random samplie. For people unschooled in sampling, however, the covariance assumption may cause problems. People may interpret a sampling procedure as a way to *cause* a distribution of outcomes (or opinions) by selecting particular traits of the population.

Study 1: A Test of the Psychological Reality of the Covariance Assumption

Method. We asked 15 sixth-grade children to design sampling methods for each of two scenarios. The fun booth scenario examined sampling ideas in the context of potential covariant relations, and the gender scenario examined students' ideas of sampling in the context of self-same relations. In the fun booth scenario, the children were told to imagine that they were going to have a booth of their choosing at a school fun fair. Everyone who went to the booth would get a prize. They had to get an estimate of how many children would come to their booth so they could estimate how many prizes to prepurchase. We told them that they were not allowed to survey everybody at the school. Instead, they could only survey 50 of the total 400 students who would go to the fair. We explained how the results of the sample could be extrapolated to estimate how many students from their school would come to the booth. In the gender scenario, we told the children that their task was to estimate how many boys and girls were at their school. As with the booth scenario, we explained (or, reexplained to the 7 students who received the gender scenario second) that they would have to take a survey of 50 students and then extrapolate to the population of 400 students at their school. Again, their task was to generate a way to choose the 50 students. For both problems, we asked the children to generate as many selection methods as they could.

If there is a covariance assumption, then we should expect the booth scenario to yield selection methods based on attributes of the population relevant to whether children would participate in the fun booth. They might, for example, sample students on a baseball team if they wanted to have a baseball toss booth. In terms of Fig. 9.1, they might want to sample the figure with the baseball hat. For the gender scenario, however, we might expect more random selection methods because the relationship between sampled entity and the property of interest is the self-same (e.g., male → male). Accordingly, we coded student responses as to whether they were based on attributes of the population that were inference relevant or whether they were random selection methods.

Results and Interpretation. There were four main types of inference-relevant selection methods. The unifying characteristic of each of these selection methods was that individuals would be selected on the basis of a hypothesis about how they would respond to the survey:

1. Self-Selection—for example, "Put up a sign, and whoever wants to come to the booth can fill it out."
2. Likely to Come—for example, "Ask all the baseball players" (for a ball toss booth).
3. Friends—for example, "I'd ask my friends, 'cause they would come."
4. Fair Split—for example, "I'd check 25 boys and 25 girls" (for the gender scenario).

We also found four main types of random selection methods. Some were more sophisticated than others, but all had the property that the selection method did not pick individuals on the basis of inference-relevant attributes:

1. Explicit Device—for example, "Draw 50 names from a hat."
2. Class/Teacher—for example, "Give it to two teachers to hand out to their classes."
3. Just Pick 'Em—for example, "You just have to pick 'em without looking."
4. First to Come By—for example, "Just stand in the hall and count the first 50 kids to come by."

There were two other types of explanation, which were not focal for this experiment, but are discussed further. The *Republic* category reflects the governmental notion that the sampled individuals can represent the opinions of others (e.g., elected officials)—for example, "I would ask 50 teachers to guess how many boys and girls there are." The *Democracy* category reflects the notion of a vote or census, rather than a sample—for example, "I'd ask everyone."

Table 9.1 shows the types of selection methods offered by each student for each scenario. Each student is represented in a vertical column. A "+" indicates a sampling method generated for the booth scenario, and an "o" indicates a sampling method generated for the gender scenario. The students are roughly ordered from left to right according to how well they fit the prediction that the booth problem would elicit inference-relevant selection methods and the gender problem would elicit random selection methods. There was a significant interaction whereby students tended to give more random selection methods for the gender scenario and more inference-relevant selection methods for the fun booth scenario: $F(1, 14) = 8.26$, $MSe = .52$, $p < .01$. Table 9.2 distills the data into a more simply understood contrast. The first column shows that 47% of the students only used random selection methods for the gender scenario: $\chi^2(1) = 4.5$, $p < .05$. The second column shows that 47% of the students only used inference-relevant selection methods for the fun booth scenario: $\chi^2(1) = 4.5$, $p < .05$. Our interpretation is that the self-same relationship in the gender scenario elicited random selection methods, whereas the potential covariance relationships in the fun booth scenario (e.g., a baseball player is more likely to come) led to selection methods based on population attributes relevant to whether the children would come to the booth.

Table 9.1 provides information relevant to the psychological complexity of a statistical inference. For the gender scenario, for example, 26% of the students suggested both a random selection method and a fair split method (i.e., ask 25 boys and 25 girls). At one moment, many of the children seemed to have a grasp of the assumptions and purposes of sampling, and at the next, they seemed to lose them, even in the noncovariant, gender scenario. In the context of the fun booth survey, children easily lost sight of the sample as a way to find information about the population at large.

TABLE 9.1

Classes of Sampling Methods Proposed by Each Student for Fun Booth (+) and Gender (o) Scenarios

	Student Sorted by Prediction Fit														
Sampling Method	*1*	*2*	*3*	*4*	*5*	*6*	*7*	*8*	*9*	*10*	*11*	*12*	*13*	*14*	*15*
Random															
Explicit Device	o							+							
Class/Teacher	o	o	o				o	o		+	+o	+o			o
Just Pick 'Em	o	o	o		o	o				o					o
First to Come By				o				o	+o		+	+			+
Inference-Relevant															
Self-Selection									+				o		
Likely to Come		+	+	+		+									
Friends	+						+o		+	+		+o	+	+	
Fair Split					o	o	o	+					o	+	
Other															
Republic												o		o	o
Democracy							+o	+							

Note. + = fun booth answer, o = gender answer.

TABLE 9.2
Percent of Students ($N = 15$) Who Proposed Random
and/or Inference-Relevant Selection Methods by Scenario

Sampling Context for Which Student Proposed Method	Selection Method	
	Random	Inference-Relevant
Fun booth only	0.0%	46.7%
Gender only	46.7%	6.7%
Both	40.0%	40.0%
Neither	13.3%	6.7%

One cause for these inconsistencies in children's reasoning may be that a statistical inference involves multiple links. Consider the middle panel of Fig. 9.1. There are at least three relationships within the statistical inference: (a) population-of-people → sample-of-people, (b) sample-of-people → sample-of-opinions, and (c) sample-of-opinions → population-of-opinions. Children who do not have a comprehensive understanding that unifies these various relations within a single structure may focus on different relationships within the inference. In the fun booth scenario, for example, there was a focus on the covariant relationship between types of people and their opinions.

THE CHARACTER OF NAIVE
STATISTICAL UNDERSTANDING

Family Resemblances to Statistical Inference Proper

As people come to understand the larger structure of a statistical inference, we propose that they rely on everyday knowledge to understand pieces of the larger structure. They draw on familiar schemas that approximate isolated relationships within the structure of a statistical inference. These schemas only have a *family resemblance* to the overall structure of a statistical inference. We use the term family resemblance because each schema shares some features with statistical inference proper, but may not share features with other applicable schemas. As we develop later, these schemas may even be incongruent with one another vis-à-vis their implications for the properties of a good sample. Figure 9.2 represents two of the schemas used for the fun booth scenario. The top half shows the notion of voting or taking a census. With this schema, the children correctly understood that the goal was to find out the population's opinion. However, the children lost the part for the whole. In importing a voting schema to understand the scenario, the children did not incorporate the fact that a sample is only part of the

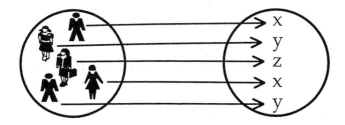

Voting/Census

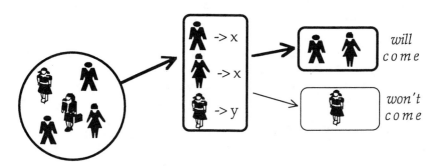

Invitation/Advertisement

FIG. 9.2. Family resemblances to a survey. A vote resembles a survey because one is finding information about the population's opinion. A vote schema, however, disregards the fact that a survey only uses a subset of the population. An invitation resembles a survey because one is addressing a subset of the population. An invitation schema, however, disregards the fact that the subset provides information about the total population.

whole population. The bottom half of the figure shows an invitation or advertisement schema. With this schema, students view sample selection as a way of soliciting or inviting a subset of people. Some of the sample will accept the invitation, as represented here by an opinion of x. Consequently, these people will be the ones who will go to the fun booth at the fair. With this schema, the students correctly recognize that sampling only involves a part of the population, but they do not consider the fact that the sample provides information about the population at large.

As we develop in the third section of this chapter, we believe that these schemas provide important initial meaning, or "prototheories," from which more complete understanding can be built. Nonetheless, these schemas can, of course, yield nonnormative notions of sampling. For example, with the fun booth scenario, 80% of the students chose to survey friends or people who were likely to come. Evidently they thought of the survey as an invitation or advertisement, and wanted to sample those people most likely to come. In the gender problem, we see hints of something resembling a fairness schema in the fact that 40% of the students thought of surveying an even split of boys and girls. In this case, they wanted to make sure that they did not bias the sample in favor of one gender or another, and evidently did not notice the undesirable side effect of their sampling rationale, namely, that it predetermines the inference about the total population.

Given the evidence of distinct selection methods, it should be evident that we do not claim that children have a singular abstract schema or a general heuristic that they use to understand all statistical situations. Instead, we view children's intuitive statistical understanding as a collection of over-lapping schemas that are differentially brought to bear depending on the particular problem context (cf. Konold, Pollatsek, Well, Lohmeier, & Lipson, 1993; Mokros & Russell, 1995). In the case of a survey, we suspect that the fact that individuals (or their attributes) cause opinions invites a covariance assumption. This elicits a particular subset of the total family resemblances on which children might rely to understand a statistical situation. Further research will be necessary to determine whether children bring a covariance assumption to statistical situations that do not involve fun fair surveys.

Different Family Resemblances Lead to Different Concerns About Sampling Violations

For the sophisticated statistical reasoner, the primary threat of violating a sampling assumption is that it leads to an unrepresentative sample. Children, however, may find different types of threats and violations that derive from the schemas that they use to make sense of the particular sampling context. Figure 9.3 portrays four characteristics of a survey and various schemas that bear a family resemblance to each characteristic. For children who focus on the size of the sample (at the right of the figure), one reason to avoid a bad (small) sample is that it may not be inclusive enough to find all the asso-ciations between people's traits and their opinions. Or, if they understand the role of sample size in terms of their prior knowledge of advertisements, a bad sample might be one that does not reach enough people. Children who focus on the part–whole (subset) relationship of a sample may rely on their prior knowledge of invitations. Consequently, a bad sample would be one that is unfair to individuals who are left out of the survey. Each schema may bring different implications for judging the quality of a sample. At the same time, some schemas may bring similar implications. For example, a sample as an advertisement and a sample as a collection of volunteers both strive for the largest sample possible.

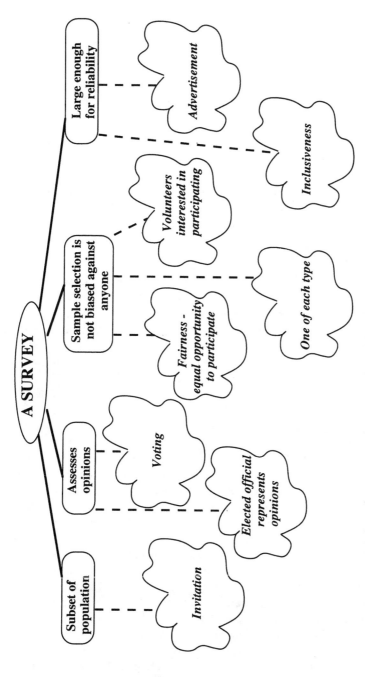

FIG. 9.3. Characteristics of a survey and possible interpretations. The rounded boxes represent properties of a survey. The clouds represent some everyday schemas that may be used to understand these properties. From Schwartz & Goldman (1996). Reprinted by permission from John Wiley & Sons, Ltd., New York.

Study 2: How Children View Violations to Their
Sampling Assumptions

To get an initial grasp on the prior knowledge and family resemblances that fifth- and sixth-grade children use to understand sampling assumptions, we conducted a second study. In the preceding study, the fact that students imported ideas of invitation and voting makes sense in that they probably did not have a strong schema for understanding the inference from sample to population in surveys. They may have "misheard" the problem and thought that their task for the fun booth scenario was to make sure that the booth had a high level of participation.

Consequently, it was important to determine how students would understand sampling when given a fuller context in which the need to make an inference to the population was emphasized.

The Task Context

To develop a fuller context for understanding sampling, we employed *The Big Splash*, one of the 12 Jasper adventures (CTGV, 1992a, 1992b). In a Jasper adventure, there is a 20-min video story during which a protagonist(s) is trying to solve a complex, real-world problem. A Jasper adventure emphasizes mathematics in one of four core domains: rate–time–distance, geometry, statistics, or algebra. At the same time, the adventures require the application and coordination of several mathematical concepts, as well as everyday knowledge, to develop a workable solution. As the video adventure develops, the problem and the relevant information unfold. At the end of the story, the protagonist faces the challenge of how to solve the problem. At this point, the students in the classroom take on the active role of trying to solve the protagonist's challenge. To do this, the students must generate, organize, and solve mathematical subgoals embedded within the video story. This is usually aided by random-access technology that allows easy revisiting to parts of the story. Ultimately, the students develop a mathematically and situationally feasible plan to resolve the challenge faced by the protagonist.

In *The Big Splash*, the main character, Chris, finds out that he could borrow a dunking machine from the fire department. He decides he would like to use the machine at the school fun fair but needs a pool to accompany the machine. He finds that there are several different ways to rent a pool and to fill it with water. Each solution involves different costs. He talks to his school's principal, who explains that she will loan him seed money, if he can show that his estimated income will be double his expenses. To get an estimate of his income, Chris takes a survey of every sixth student in the school's cafeteria line. He asks the students whether they would come to the dunking booth, and if so, how much they would be willing to pay for a chance to dunk a teacher. Within the video, there are further subproblems including how to make sure the booth and pool will be ready in time for the fair. For our purposes, however, the most relevant portion of this Jasper adventure involves how children think about taking a survey to get an income estimate.

Method

As is common with Jasper adventures, the 45 fifth- and sixth-grade students worked a full week, 1 hour per day, developing their plan for Chris. Of relevance to the issue of sampling, the students estimated Chris's projected income by extrapolating from the results of his survey. We thought that this activity and the surrounding context would ground their understanding that the purpose of a sample is to infer a population distribution. After students completed *The Big Splash*, we assessed the students' understanding of sampling.

We used paper-and-pencil instruments to capture the students' understanding of sampling in both survey and chance setups. For the survey setup, we developed a textual story about Amy. In the story, Amy has decided to take a survey about her fishing-for-prizes booth by asking every fifth student in the school cafeteria line. The story explains that the children line up for the cafeteria with their classes of 20 students. It also explains that there are four grades ranging from first to fourth. Thus, the story includes information relevant to collecting a representative sample.

After reading and hearing the cover story, students considered different ways that Amy could have taken her survey. The students received forced-choice questions from which they had to choose their preferred method of sampling—for example, "Which is the better way for Amy to conduct her survey: (a) Ask ½ the children in each grade, or (b) Ask all the children in the 2nd and 4th grades." Afterward, the students had to write what was wrong with the method they did not choose. (This turned out to be a very productive method of eliciting students' thinking, at least compared to "Explain why.") The forced-choice items presented alternative sampling methods that covered the issues of biased sampling methods, sample size, stratification (without any mention of randomization), and three types of random selection method, which we label random device (draw from a hat), stratified random (sample based on order in the line), and random preselection (sample on basis of preexisting, but inference-irrelevant, attributes). Throughout the chapter, we distinguish between questions that described stratified sampling methods and stratified-random sampling methods. The difference has to do with the amount of information included in the question. We refer to a "stratified" sampling method when there is no mention of any randomizing procedure—for example, "Ask ½ the children in each grade." We refer to a "stratified-random" sampling method when the randomization procedure is included, however obliquely, in the sampling description—for example, "Ask every 10th student in the cafeteria line." This method is stratified because the students line up by class, and it is random because it selects every 10th student in line. We made the distinction between stratified and stratified-random samples because we thought it would impact the students' reasoning.

In some of the forced-choice items, we crossed several different sampling issues. For example, one item asked if it were better to ask the 74 students whose last names begin with A to H, or to ask a small sample of 40 students by randomly choosing two children in each class. In the first choice, there is a relatively large sample that is chosen on the basis of preexisting attributes (i.e., names). In the second choice, there is a relatively small sample that has been selected with a stratified-random procedure. Crossing issues allows us to examine the relative contribution of different beliefs about sampling. For example, children may be willing to give up a large sample to avoid a preselection sampling technique.

To evaluate the students' grasp of the relationship between a sample and a population in a chance setup, we administered multiple-choice questions using dice and urn problems. We were particularly interested to see if the students could use a sample to draw inferences about a population, and whether they could predict samples from a population in a chance setup. For example, one question read as follows: "Imagine a bag of marbles has 60 white marbles, 20 black marbles, and 20 red marbles. If you pulled out 10 marbles, what do you think you would get: (a) 5 whites, 3 blacks, 2 reds; (b) 6 whites, 2 blacks, 2 red," and so forth. We also included sample/population questions in the context of dice: for example, "Imagine a die with 5 black sides and 1 white side. What do you think you would get on 6 rolls: (a) 6 blacks, (b) 5 blacks and 1 white, (c) 4 blacks and 2 whites," and so forth. We did this for two reasons. First, when using a die, the population of six sides can be smaller than the sample size (e.g., 60 rolls), thereby constraining estimates of the population (e.g., 6 blacks, 5 blacks and 1 white, etc.). This is the inverse of the urn situation where the estimates of the sample are more constrained (e.g., 10 blues, 9 blues and 1 white, etc.). We wanted to find out if this difference between the dice and marble contexts would influence children's ability to infer a population from a sample. Second, dice make it relatively easy determine whether the saliency of a distribution has an influence on reasoning. In particular, we were interested in finding out whether children would reason about 60 rolls the same way they would reason about 6 rolls of a die. If they do not, this would show that even the specific context of a chance setup can influence children's reasoning.

Results and Interpretations

For Chance Setups. First, we consider what students understood about samples and populations in chance setups. Table 9.3 shows the percentages of correct responses (out of the seven possibilities for each question). These

TABLE 9.3
Percent of Students ($N = 45$) Who Chose the
Most Likely Alternative (Out of 7) in a Chance Setup

	Population = 100 Sample = 10 Marbles	Population = 6 Sample = 6 Dice	Population = 6 Sample = 60 Dice
Given sample, infer population	60%	86%	81%
Given population, infer sample	62%	71%	71%

results indicate that the majority of students were able to match samples and populations. It seems, however, that children were primarily matching sample and population ratios (e.g., 70:30 population and 7:3 sample) rather than reasoning about probabilities in a normative sense. Most of the students who did not match proportional samples and populations selected alternatives that were "off" by one or two. These students were probably reasoning that getting a perfect match would be too lucky. The students did not reason that chance makes it so a sample that is proportional to a population is the most likely of all the possible sample distributions. Instead, chance explained why one would not get an exactly proportional or representative sample. Consequently, they chose the off-by-one alternatives to symbolize the unlikeliness of getting an exact match (Bar-Hillel, 1980). Supporting this idea, one may notice the high accuracy rates on the die questions that required inferring the population. Because a die only has 6 sides, students could not choose an off-by-one answer without seriously mismatching ratios; given 3 black and 3 white roles, a 4 to 2 ratio on the die would be a very nonproportional estimate. All told, these results suggest that students believe that samples and populations will have similar distributions but that chance makes it unlikely that one would get a perfectly representative sample from a population. The truth is that a sample that is proportional to the population is the most likely outcome, even though it may be unlikely in an absolute sense.

It is informative to note that even in the chance setup, there was evidence of an effect of context on statistical reasoning. This evidence is best understood by considering the failure to replicate Konold's (1989) work demonstrating an outcome-oriented approach to estimating samples. Using an outcome-oriented approach, people focus on the likelihood of a single event (e.g., a black roll). They do not reason about the likelihood of a distribution of events (e.g., five blacks and one white). When they are asked to reason about a distribution of events, such as what is the most likely distribution given six rolls, they tend to extrapolate the most likely result for a single event. For example, given a die with five black sides and one white side, a single roll should yield a black. Consequently, they might incorrectly infer that six rolls should lead to six black outcomes, one on each trial. Konold (1989) presented supporting evidence for an outcome-oriented approach by showing that 60% of adult subjects predicted that six rolls would yield six blacks for a die with five black sides and one white side. Our students failed to replicate this result. Only 7% of the students chose six black outcomes, whereas 71% chose the correct five to one alternative. The difference between our results and Konold's may be due to the fact that our students saw several possible distributions as a result of the multiple choice format (e.g., 6:0, 5:1, etc.). This may have led them away from the univocal outcome-oriented heuristic toward a more distributional focus (cf. Konold et al., 1993). Interestingly, for the problem with 60 rolls of the die, 0% of the students in the current study chose 60 black outcomes. Thus, as the number of rolls became large, there was a complete shift away from an outcome-oriented approach. One interpretation is that people do not use an outcome-oriented approach for larger samples for which reasoning about distributions becomes more worthwhile. Fitting our overall story, people's statistical reasoning varies depending on the case at hand.

For Survey Setups.   Next we consider the fifth- and sixth-grader's notions of how to generate a good sample in the context of a survey. Table 9.4 shows

TABLE 9.4
Questions and Choice Percents for Amy Sampling Questions ($N = 45$)

% Sampling Method	Choice of Sampling Methods
3% Biased	Give the surveys to all students in Grades 1 and 3.
97% Stratified	Give the surveys to half the students in Grades 1 thru 4.
21% Biased	Ask the first 20 and last 20 students in the cafeteria line.
79% Stratified-random	Ask every 10th student in the cafeteria line.
31% Small sample	Give a survey to 20 students in the cafeteria line.
69% Large sample	Give a survey to 70 students in the cafeteria line.
76% Large sample	Ask 80 students by asking every 5th person in cafeteria line.
24% Huge sample	Ask 200 students by asking every 2nd person in cafeteria line.
59% Biased	Ask the first 60 students in the cafeteria line.
41% Random device	Put all the students' names into a hat, shake it up, pull out 60 names and give the survey to those 60 students.
16% Random preselect	Ask the 64 students whose phone numbers end with the number 3.
84% Random device	Put all the students' names in a paper bag, mix the names, pull out 64 names, and ask those students.
34% Large random preselect	Ask 76 students whose last names start with the letters A, B, C, D, E, F, G, H.
66% Small stratified-random	Ask 40 students by going to each of the school's 20 classrooms and randomly picking 2 students.

Note. Background information for sampling scenario (see text for further details): There are 400 students at Amy's school in Grades 1, 2, 3 and 4. There are 100 students in each grade and five classes in each grade. At lunchtime, all the students in the school line up in the cafeteria line with their classes.

the percent of students who preferred one or the other sampling method in each pair. First we consider evidence for ideas that could be built into fuller understanding; then we look at evidence of incomplete knowledge.

The students preferred stratified samples that did not mention a randomization technique over biased samples. Ninety-seven percent would ask half the students in each grade over asking all the students in two of the four grades. They also preferred stratified-random samples over biased samples. Seventy-nine percent of the students would ask every 10th student in the line over asking the first and last 20 students in line. (Later, we suggest why students did not prefer the stratified-random sample as much as they preferred the stratified sample.) The students also hinted at the central limit theorem by preferring a survey size of 70 over 20, and a survey of 80 over 200.

Their justifications for why they did not like some sampling methods also suggest reasonable conceptions. Consider the item that asked whether Amy should give out the survey (a) to the first and last 20 students in the cafeteria line, or (b) to every 10th student in the cafeteria line. Thirty-eight percent of the students rejected the biased alternative (a) with the justification that it was either unfair or had less variety of opinions represented. For example, one student wrote, "You would only have 2 grades, instead of a variety of grades," and another wrote, "You have to give the others a chance. She has to spread them out." Thirteen percent of the students rejected the biased sample because it violated independence. They wrote that a child could influence the opinion of another child in the line. Thirty-six percent of the students approximated the law of large numbers by claiming that one sampling method had too few kids compared to the other. This subset of the students primarily chose the biased alternative because they miscalculated how many children Amy would sample when asking every 10th student in line. The remaining 13% either felt that the nonbiased method was too much work (10%) or wrote undecipherable justifications (3%). Thus, even though only 79% chose the stratified-random sample, 87% of the students gave justifications that had family resemblances to normative considerations in taking good samples.

Despite these promising intuitions, the students did not have an articulated concept of sampling as it relates to the full structure of a statistical inference. For example, the students disliked biased samples, as shown by the 97% rejection rate of a biased sample when it was pitted against a stratified sample. Yet 59% of the students preferred a biased sampling method that selected the first 60 children in line over a random method that drew 60 names from a hat. Many students' beliefs about drawing names from a hat are captured in the explanations of two students. One wrote, "She might pull out all first grade names," and the other wrote, "[It] is too hard and is stupid. You want random kids." A large number of children accepted a biased sample to avoid using a randomizing device in the context of a survey. We also see hints of this avoidance of a randomizing scheme in the responses to the stratified-random method that asked every 10th child in the cafeteria line. Although the randomizing scheme was subtle, the students were less inclined to accept the stratified-random method (79%) relative to a stratified sample that had no mention of randomization whatsoever (97%). Unfortunately, we did not pit the two stratification methods against the same biased sampling method, so our interpretation is necessarily speculative.

If students are disinclined to use a random sample in a survey setup, how do they conceive of a good sample? According to the covariance assumption, the children unduly focus on characteristics within the population. If the children are offered any means of characterizing the types of people that constitute a population, they want to make sure that each type of person is included in the sample (e.g., to be fair, to get one of each type, etc.). Thus, children may not have liked the idea of a random sample because it left the selection of population characteristics up to chance. It is much better to have a stratified sample in which each type of individual is included in the sample. As evidence to this point, consider two findings. First, even though 59% of the students did not like the idea of using a random draw when compared to a known biased sample, 84% preferred drawing 64 names out of a hat over choosing the 64 students whose phone numbers end with a 3. Second, even though 69% of the students preferred a large sample to a small one, 66% preferred to take a small sample that drew from each class over taking a large sample that was based on the first letters of last names. In each case, the nonpreferred option excluded a segment of the population from possibly participating in the sample. These results may explain why students disliked biased and small samples. They preferred sampling methods that they thought increased their chances that all kinds of people would be included.

This study shows that many students had a grasp of making inferences between samples and populations in chance setups. However, when sampling was placed in the context of a survey, the children imported schemas that influenced their assumptions about what makes a good sample. We found a loose ranking in selection preferences, with stratified methods being the most preferred, stratified-random next, then randomizing devices, and finally, selection methods that depended on preexisting conditions being the least preferred. The rejection of preselection criteria, like the last digit of a phone number, can have a normative basis; for example, the phone company may give wealthy people phone numbers that end with a 3. Perhaps students did not view phone numbers as inference irrelevant. Maybe they thought that phone numbers could be systematically assigned to people's opinions about the fun fair booth, and therefore, preselection based on this attribute would have biased the results. On the basis of this experiment, we do not know if the students disliked preselected samples because they might bias the results. However, given the students' overwhelming preference for samples that included some form of stratification, we suspect that they rejected preselected samples, in part, because they excluded a portion of the population with identifiable characteristics. This makes sense if one construes the purpose of a sample as identifying the covariances between population characteristics and opinions, or of making sure that every type of person is fairly included in the sample. We examine the basis for the students' rejection of preselected samples more closely in the next study.

BUILDING ON COMMON SENSE: ANCHORED
INSTRUCTION AND ASSESSMENT

It seems clear from the two studies already reported that children rely on prior knowledge of related events to make sense of sampling. We believe it is important to build on the initial understandings that students use to make sense of different sampling situations. Otherwise, the students may learn statistical concepts that they can only apply in chance setups. To emphasize the value of commonsense ideas as stepping stones to fuller understanding, we call them *prototheories*. A prototheory is based on a schema of situations that bear a family resemblance to sampling. It is a partial and unratified understanding of sampling that works in some situations but not others. A prototheory provides meaning that needs to be sharpened, articulated and aligned with other prototheories to yield fuller understanding. For example, the prototheory that good samples should be stratified is a powerful idea frequently used in experimental research. Similarly, for the chance setups, the children's belief that randomness makes a perfectly representative sample unlikely may serve as a prototheory for understanding that a representative sample is still the most likely outcome. Yet, in their prototheory forms, these two ideas may be incompatible. Even though the children could grasp stratifying in a survey setup and randomness in a chance setup, they did not seem to have a grasp of the rationale for taking a random sample in a survey setup. They had not realized that one takes a random sample precisely because one cannot identify and stratify all the population traits that might covary with different opinions. Consequently, it is important to allow children to align their prototheory of random sampling in chance setups with their prototheory of stratification in survey setups. As pointed out by diSessa (1993) in the domain of physics, the goal of instruction is to provide a means by which "knowledge in pieces" can be fit together to yield a more coherent body of knowledge. Conceivably, in the domain of statistical inference, the attempt to reconcile the two prototheories could yield an advancement in student understanding.

The Instructional Challenge of Aligning Prototheories of Sampling

Our way of aligning students' prototheories into more articulated structures is to provide instruction that brings multiple prototheories into juxtaposition. If we assume that each prototheory is based on different family resemblances brought to mind by a particular context or focus, then it is possible for children to slip between prototheories without considering the relationships between them (refer to Fig. 9.3 for examples of schemas that may serve as prototheories). For example, one might think in terms of social fairness when considering a sample one way (e.g., an election), and then one might think about trying to infer the covariant relationship between people's traits and opinions when thinking about the sampling situation another way (e.g., an experiment). Because the application of each prototheory may result from a focus on different aspects of a statistical inference, it is possible that the two understandings would never be simultaneously available to constrain one another. Even worse, one prototheory may inhibit the application of another as in the case of a causal focus suppressing a statistical focus. Finally, children may not recognize the similarities between sampling techniques usually reserved for different contexts. For example, although the children had little difficulty with the idea of drawing marbles from an urn, they may have rejected drawing names from a hat in the survey context because they had no exposure to randomizing devices in surveys. The instructional challenge of the incongruency between everyday and mathematical understanding is that everyday understanding is highly applicable and meaningful, yet often disconnected and subtly different from more general and normative mathematical understandings. Because the child can move from one prototheory to another, within and between contexts, the challenge is to pin down the prototheories so they may inform one another.

A Model of Anchored Instruction and Assessment to Align Common Sense

To help children synthesize their different prototheories, we rely on anchored instruction in which a single problem-solving macrocontext serves as the situational crucible where different prototheories are brought to bear, juxtaposed, and hopefully articulated and aligned. As part of a larger effort to help students make connections between, and elaborate on, their mathematical and contextual understandings, we are developing a new model of assessment (Barron et al., 1995). The model that we are exploring to help children synthesize their different prototheories is depicted in Fig. 9.4. In this model, hereafter SMART (Scientific and Mathematical Arenas for Refining

The SMART Model

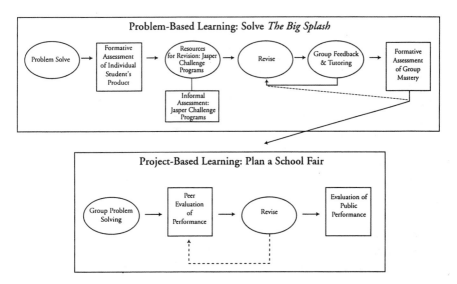

FIG. 9.4. A model of integrated instruction and formative assessment.

Thinking), students first explore target concepts in the context of a complex problem or "anchor" and then extend and deepen their understanding in the context of a related project. In a recent study, for example, students solved a Jasper problem called *Blueprint for Success,* and then designed playhouses that were given to local organizations. Culminating projects, in addition to their motivational and rehearsal value, allow for further context-sensitive ideas to reveal themselves as children bump up against the world in their planning. In this chapter, however, we focus on the initial problem-based portion of the model (see Barron et al., 1995, for a description of the complete model).

A central assumption of the model is that the processes of objectifying thought and of negotiating meaning with one's peers and more knowledge-able others are fundamental to conceptual evolution (Cobb et al., 1992; Schwartz, 1995). We rely on the anchor problem to create a shared context in which students' prototheories can be made "visible" to teachers and other students through discussion and in the form of concrete products. Hopefully, once visible, these ideas will be juxtaposed, reflected on, and aligned. From our perspective, this process of reflection is a process of formative assess-ment, and for this reason we conceptualize SMART as a model of integrated instruction and assessment. The model incorporates specific features, which we discuss next, that are designed to further support and formalize the assessment process.

Cycles of Work and Revision

In SMART, instruction is organized around iterative cycles of problem solving and revision. Students generate an initial solution to a circumscribed portion of the anchor problem and then redraft their solutions after accessing a variety of information resources. In this way, the instructional sequence supports both the articulation of prototheories and their reformulation. Although reformulation may seem to be an obvious precondition for conceptual evolution, our work in classrooms has shown that students very rarely have opportunities to revise their work. More typically, students receive instruction, complete a related assignment for which they may or may not get a grade, and then move on to a new topic of instruction. In contrast, SMART students expect to have opportunities to reflect on their "first drafts" and to have access to resources that will help them rethink concepts and improve their work.

Programs That Make Thinking Visible: Video Resources for Discussion and Reflection

One way to help conceptual growth is to make prototheories explicit and the focus of discussion. Discussion provides opportunities for students to reflect, and for the teacher to evaluate students' underlying conceptual models, and tailor instruction accordingly. One resource that we developed to facilitate dialogue is a series of supplementary videos called the Jasper Challenge Programs. There are four programs that accompany *The Big Splash*; each program focuses on a set of concepts pertinent to *The Big Splash* solution. Programs are shown on a just-in-time basis, so that the concepts relate to the subproblem that the students are currently solving (e.g., extrapolating from a sample). The programs are used in a whole-class format that, as Fig. 9.4 shows, occurs preparatory to revision. The programs are also used during revision by individual students who review different segments according to their individual needs.

Each Challenge Program has four segments: Toolbox, Smart Lab, Kids Online, and The Challenge for Next Time. The Toolbox segment is hosted by Dave, who specializes in generating visual representations that aid in problem solving. Dave provides ideas for visual representations that can be conceived of as "tools" for problem solving and communicating. We chose visual representations for a number of reasons, including their usefulness for revealing patterns and communicating mathematical ideas in a nonprocedural way. Toolbox was not designed to give away solutions, but rather to provide scaffolding for the articulation and integration of everyday and mathematical knowledge. There were two Toolbox segments that bear directly on sampling issues.

One Toolbox segment was directed at the covariance assumption that students use to link people and their opinions in a survey. Dave frames the survey in a chance setup (see the bottom panel of Fig. 9.1). He shows a large jar of colored chips. He explains that each color represents a different student opinion about the dunking booth. For example, yellow equals a student who would pay $.50, red equals a student who would pay $1.00, and so forth. He then draws several samples of 60 and 15 each. Afterward, he shows how to organize the results into a histogram and shows how the estimates resulting from each sample of 60 are quite similar, but that the samples of 15 yield more variability. We thought that the explicit use of a chance setup sampling technique within the survey context would help students align their notions of sampling in the two situations. In particular, they might reconcile the random selection they accepted in a chance setup with the stratified selection they embraced in the survey setup.

In a second relevant Toolbox segment, we targeted the inferential link from sample to population. Dave shows how one can extrapolate from a sample to find an estimate of the population. He does this by creating a box to represent the 60 students who were sampled. A shaded part within the box represents the students who said they would dunk the teacher, and the unshaded part represents the students who would not. Dave shows that six of these boxes would represent the school population of 360 students. He draws the remaining boxes and invites the children to figure out how many students would be in the shaded region. We thought this demonstration of extrapolation was important for two reasons. First, students of this age do not have a solid grasp of proportional relationships, and providing them with this method of additive extrapolation might help them think about the population and sample ratios (Moore & Schwartz, 1994). Second, the demonstration emphasizes that the sample is only part of the population, and that the sample is used to provide information about the total population. Our hope was that this Toolbox segment could help students integrate their prototheories about good samples around the larger purpose of a statistical inference. For example, with an understanding of this estimation purpose, students might be able to infer that a bad sample is one that yields a biased estimate of the total population.

The Challenge Program segments Smart Lab and Kids Online are similarly designed to promote discussion and informal assessment. Smart Lab presents data in graphic form on how a sample of students thought about or solved some aspect of *The Big Splash*. For example, one Smart Lab contains a graph showing how the sample group answered the question, "What is a representative sample?" By collecting Smart Lab data from their own class, teachers are able to assess students' understanding and, at the same time, create natural opportunities to discuss and negotiate different conceptions. In Kids Online, student actors present their ideas and solutions in detail. For example, in one segment, the presenter argues that a representative sample is one where the opinions of the sample are like a "miniature version" of the opinions in the population. The presenter goes on to describe an experiment he conducted that shows what can happen if this condition is violated. The task of the student-viewer is to evaluate the presenter's thinking critically. By design, Kids Online has both presenters whose conceptions and reasoning are sound, and those whose thinking is faulty. We deliberately included presenters who articulate prototheories that are characteristic of real students so as to ensure that these prototheories become controversial.

The fourth assessment tool, The Challenge for Next Time, is designed to help teachers probe students' understanding of key concepts. A "challenge" question appears after a program segment has been viewed. The challenge pertains to the information in the segment. For the Toolbox segment on extrapolation, the challenge question asks students to imagine a new sample size, and to draw a new representation showing the relationship of the sample to the school population. Students individually write responses to the challenges, providing teachers with additional assessment information.

Rubrics for Assessment and Feedback

Early in our work on SMART, teachers expressed an interest in ways to evaluate student work formally. They reported that the information they collected in the context of the Jasper Challenge Programs was useful, but was transitory and often incomplete. At the end of a lesson they did not have always have a picture of what each child in the class understood, or how a given child's understanding might have changed over the course of a lesson. Also, and perhaps not surprisingly, they reported that they did not know how to evaluate the complex products generated by students. In response, we developed formal evaluation procedures that are formative in nature and functional for both teachers and students. Specifically, after students draft their solutions (see Fig. 9.4), teachers evaluate them and provide written feedback. Teachers use a computer-based checklist to evaluate aspects of students' work. The teacher clicks on the relevant feedback statement and may add a comment if desired. When the teacher finishes evaluating a student's work, an individualized report is printed. This report contains revision-relevant feedback, and also some suggestions for information resources that may be helpful to the student. For example, the report may suggest that the student review a specific segment of the Jasper Challenge Program, review a "model" solution to a near transfer problem, review relevant segments from *The Big Splash*, or consult with peers.

There are several important aspects of this evaluation tool. First, teachers use it just prior to showing the Jasper Challenge Programs, and in this way have information they can use to better tailor the discussion to address particular conceptual issues that the class may have. Further, the evaluation is sequenced such that the resulting feedback serves as a resource for students during the revision phase of the instruction. Second, the feedback is relatively general and neither provides answers nor a set of procedures to follow. Instead, the feedback orients students to rethink specific issues. Finally, consistent with our emphasis on mathematics concepts, the foci of the evaluation are concepts, and not rote procedures.

Study 3: The Effects of Situated Instruction and Assessment on Prototheory Alignment

Method. In the current study, we wanted before and after snapshots of student understanding to see how it evolved as a result of working on *The Big Splash* and the Jasper Challenge Programs. In the study, 123 fifth graders answered a number of paper-and-pencil problems relevant to sampling methods before and after instruction. During instruction, the students solved *The Big Splash* with the benefit of the Jasper Challenge Programs. Because the Challenge Programs engaged students and teachers in numerous activities that embellished mathematical concepts, mathematical practices, and conceptions of assessment as a formative process, the intervention took approximately 2 months to complete. In addition to the sampling-relevant activities mentioned earlier, the students worked on other concepts and skills including rates, cumulative frequencies, net profit, planning, and justifications. Consequently, there were numerous questions associated with these concepts on the pre- and posttests. However, for the current purpose, we extract the questions relevant to sampling (see Table 9.5).

We included the subset of the forced-choice Amy questions from Study 2 that pitted biased selection methods against stratified-random and random-device selection methods. This allowed us to see whether students had prototheories about what constitutes a good sample, whether they evolve, and whether our intervention helped students embrace both random and stratified-random selection methods for the survey context.

We also included the open-ended question, "How could Amy use a survey to help develop her plan?" This question, coupled with the justifications for the forced-choice questions, allowed us to see whether students understood that a sample was different from the population, and whether the students understood that a sample provides information about the population. For example, if students wrote about the survey in terms of an invitation, then they did not receive credit for understanding that a sample provides information about the population. If students noted that the sample could be used to estimate the total number of students who might come to the booth, then they received credit, assuming that they did not show evidence of an alternative interpretation elsewhere. This coding system is conservative in the case where a child shows an understanding that a sample informs about a population on one answer but shows an invitation schema on another answer. In this case, we coded the child as not showing evidence of understanding. We also did not give credit to children who did not articulate their reasoning one way or another.

Finally, there were two multiple-choice problems. In one, we asked the students what they thought about a survey that selected the 76 children whose last names started with the letters A through M. The choices were whether the sample was (a) OK, (b) too small, (c) bad because it was unfair to N–Z children, or (d) bad because N–Z children have different opinions than A–M children. This question allowed us to see if children rejected preselected samples for the normative reason that they may bias the estimate (d), or for the prototheory-based reason that they are unfair to those who are excluded from the sample (c). It also allowed us to see if students' evaluation of preselection sampling methods changed in concert with their changing views on random and stratified selection methods. The second multiple-choice question asked about Amy's estimated income if she sampled 50 of her friends: Would it be too high, too low, just right, or is it impossible to tell? The children also justified their choices to this item in writing. We thought that this question would further illuminate whether students understand that a biased sample can influence one's estimate of the population parameters. Presumably, responses to this question would be contingent on the children's understanding that a sample provides information about a population, an understanding that we evaluated from the open-ended question.

Results and Interpretations

Purpose of a Survey.   Table 9.5 shows the response percents on the pre- and posttests. The two items represented at the top of the table, items 1 and 2, show that prior to the intervention, many students did not view a survey as a tool that can reveal properties of a population. Instead, they tended to interpret the survey by way of its family resemblances to an advertisement, a method for selling tickets, an invitation, a vote, a way to get second opinions, or a raffle. Each of these interpretations misses one or both of the part–whole and informational features of a survey. For example, the vote interpretation of a survey means that students had encoded the informational purpose of the survey but not the part–whole relationship. An invitation interpretation means that students captured the part–whole nature of a survey but not the informational purpose. After the intervention, Amy's survey was more frequently viewed as a method for determining the characteristics of the complete population. This result suggests an important reason that contextual instruction is important. Real-world contexts help students interpret the world in mathematical terms—in this case, in terms of the relationships that allow extrapolating from the measured sample to the whole population.

TABLE 9.5
Pre–Post Percents for Study 3 ($N = 123$)

	Pre		Post		
	Yes	No	Yes	No	Students Exhibited Knowledge That:
(1)	49%	51%	81%	19%	Sample is different from total population that could attend booth.
(2)	26%	74%	65%	35%	Sample provides information about total population.

(3) What if Amy took a survey of 50 of her friends? Her estimate of how many tickets she would sell would probably be:

Pre	Post	
16%	31%	Too high
7%	9%	Too low
35%	20%	Just right
42%	40%	Can't tell

	Pre		Post		
	Yes	No	Yes	No	Students Preferred Normative Sampling Method:
(4)	42%	58%	67%	33%	Stratified-random over biased (every 10th vs. first/last 20 in line)
(5)	41%	59%	61%	39%	Random device over biased (hat draw vs. first 60 in line)

(6) What if Amy took a survey of the 76 students whose last names start with A–M?

Pre	Post	
14%	18%	Sample is OK.
13%	16%	Sample is too small.
35%	47%	Sample in unfair to N–Z students.
38%	19%	Sample gives bad estimate because N–Z students have different opinions.

Note. Refer to text for explanation.

Understanding of Bias. Given the improved understanding of the link between sample and population, we can ask whether it was accompanied by an improved understanding of why a biased sample might threaten the validity of one's conclusion. Item 3 in Table 9.5 provides the relevant information. At pretest 35% of the students thought that sampling Amy's friends would give just the right estimate, whereas only 16% felt it would give an overestimate. Many of the justifications for why it would give a good sample reflected an invitation schema. For example, one student wrote, "Because the kids all like her, they will all come." In this case, the sample and the subset of the population who will go to the booth have not been differentiated. On the posttest, 20% of the students felt that surveying friends would give a good estimate, whereas 31% felt that it would yield an overestimate.

Although the 15% swing from pre- to posttest is encouraging, it is not as large as we had hoped. One reason that the switch may not have been larger is that students brought many ideas to this question that were not adequately discriminated by the multiple-choice format. For example, about 40% of the students chose the "Can't tell" option at pretest. When examining the written justifications of those students who chose the "Can't tell" option on the pretest, 33% justified their choice by explaining that they could not make a decision without knowing more properties of the sample—for example, "I don't know who these people are so I can't say," and "depends on the ticket price." In this case, the students were thinking in terms of the covariance relationship between people (and ticket) characteristics and the sampled individuals' willingness to pay to dunk a teacher. In contrast, only 14% of the students who chose the "Can't tell" option on the posttest offered this form of justification. Interestingly, several subjects justified their "Can't tell" and "Too low" choices on the posttest with explanations like, "Just because they all like her doesn't mean they'll buy a ticket," or, "Because her friends would be cheap and put a low price." For the researcher well versed in threats to sample validity, there is a tacit understanding that when answering this question one should assume that friends will pay more; otherwise, why would one have a question like this? However, not all the students construed the question with the set we had in mind when we constructed the problem.

The current study cannot demonstrate that understanding the part–whole relationship between sample and population caused a better understanding of biased estimates. However, the part–whole relationship is implicated in the understanding of bias. Of those students who learned from pre- to posttest that sampling one's friends would yield too high an estimate, 35% had learned that a sample is part of the whole, and 55% knew the part–whole relationship at both pre- and posttest as shown by questions 1 and 2. Evidently, the part-whole relationship plays an important role in understanding bias, as only 10% who improved on the "Amy friends" question did not understand the part–whole relationship. However, understanding the part–whole relationship is not sufficient for understanding bias. This is demonstrated by the fact that a large percent of the students who improved on the bias question already knew the part–whole relationship at pretest (i.e., 55%). This highlights the numerous understandings, of which the part–whole relationship is one, that must be brought to bear to understand the assumptions behind a valid statistical inference.

Randomization and Stratification. Items 4 and 5 reveal students' beliefs about stratified-random samples and samples using a randomizing device. At pretest the majority of students did not like either stratified-random samples or random-device samples. The posttest percents show that the intervention was successful at improving students' understanding with respect to both sampling techniques. The students probably did not reach the level of performance found in the previous study because of a number of different experimental procedures, including the use of younger and less accomplished fifth graders, a briefer response period for each question, and a numbingly large number of test items that embraced the full instructional range of *The Big Splash* and the Challenge Programs.

An interesting question is why the students at pretest, presumably operating under the covariance assumption, were not more favorable toward the stratified-random sample. One possible explanation is that the students had minimal understanding of the purpose of the survey to start with. Another possibility is that the students had never seen an example of a stratified sample that systematically pulled random students from a line. Unlike the prior study, these students were responding prior to seeing *The Big Splash* with its example of this form of sampling. Consequently, pulling people from a line may have seemed like an odd and effortful proposal, perhaps as odd and effortful as pulling names from a hat.

Reconciling Prototheories Within a Fairness Schema. In study 2, we pointed to the children's preference for a stratified sample in a survey context. They want their sample to represent each segment of the population. This may be called a stratification prototheory. At the same time, we found a prototheory about random samples in chance setups. Children believe that the chance elements within a random sample will make it unlikely that one will get a perfectly representative sample. These two prototheories may be incongruent with one another in the context of a survey. The children want to get one of each type of person (stratify prototheory), but a randomizing device does not ensure such a selection (chance prototheory). The current study also found evidence of this incongruence, as only 19% of the students accepted both stratified-random and random-device selection methods on the pretest.

Perhaps as a result of the Toolbox segment that used a random draw of colored chips to model a survey, 44% of the students accepted both random-device and stratified-random methods at posttest. An interesting question is how children reconciled their stratification and randomization prototheories. Item 6, which tests children's view of preselected samples, provides one possible clue. In this item, we tested two possible reasons that students did not like samples that selected individuals according to preselected, but most likely inference-irrelevant, dimensions. One reason was that it was unfair to the students left out of the survey. The other reason was that it led to a bad estimate because the unselected population may have systematically different opinions than those included in the sample. On the pretest, there was a fairly even, and perhaps chance based, split between these two reasons for rejecting the preselected sample. However, on the posttest there was an interpretable shift with 47% of the students choosing unfairness as a reason to reject a preselected sample. One possible explanation for this shift has to do with how children came to accept a random sample in a survey context. In their attempt to reconcile their stratification and chance prototheories in the survey context, they may have relied on a fairness prototheory. A fairness prototheory might state that a good sample does not exclude any segment of the population from having a chance of being in the sample.

Support for the interpretation that the reconciliation of the two prototheories was associated with the fairness schema would be found if students who accepted both sampling methods tended to pick the fairness rationale for rejecting a preselected sample. Out of the 44% of the students who embraced both stratified-random and random-device samples at posttest, 50% chose the fairness option from item 6. For the students who only embraced one of the sampling methods at posttest, 35% chose the fairness option.[1] These results tentatively suggest that a fairness schema provided a means for aligning the stratification and chance prototheories.

The fairness schema represents an advancement and alignment in the students' understanding of the originally incongruent notions of stratification and chance. Fairness is an excellent heuristic for evaluating the quality of a selection method and covers a large range of situations. The fairness prototheory, however, does not completely abandon the covariance assumption. Students still do not think in terms of what is being sampled; rather, they think in terms of what is generating or causing the sampled outcome. Consequently, a fairness prototheory does not support an evaluation of a selection method on the basis of its potential effect on the inference from sample to population. So, for example, we predict that students armed with a fairness prototheory would not spontaneously recognize the problem of self-selection endemic to phone-in surveys (cf. Jacobs, 1996). This is because this sampling method is fair in that everybody has an opportunity to call and voice their opinions if they want. The fact that people with one opinion or another may be more likely to call is not within the inferential sphere of the fairness prototheory.

A Note on Between-Subjects Variability. We have tried to indicate the general shifts in the children's understanding as a group. We have also tried to document some of the shifts within individuals by showing that a

---

[1]Of the students who only accepted either stratified-random or random-device selection methods at pretest, 34% chose the fairness option. Of the 19% of the students who embraced both stratified-random and random-device sampling methods at pretest, only 23% chose the fairness option. This latter result may indicate that some students had a fuller understanding of a statistical inference at pretest.

grasp of the part–whole relationship accompanied a grasp of sampling bias, and by showing that students who accepted both stratified-random and random-device samples tended to use fairness as a basis for evaluating the quality of a sample. It should be noted, however, that very few students developed an understanding of the full structure of a statistical inference as it relates to sampling assumptions. On the pretest, no students were normative across all the measures, and at posttest only 5% of the students were consistently normative. To some extent this may be attributable to imperfect items that allowed students to give nonnormative responses for hidden yet normative reasons. Nonetheless, the lack of normative consistency within individuals indicates that it was not the same set of students who caused the pre- to posttest gains on each item. We believe this lack of consistency accurately reflects students' understanding of sampling at this age. The children have a collection of schemas that they differentially bring to bear depending on the personal saliency of particular aspects of each question. Some of these nascent understandings become reconciled during the lesson and represent an advancement in the children's knowledge, but others remain detached. This should not be surprising, given the complex structure of a statistical inference.

SUMMARY AND CONCLUSIONS

In this chapter we examined fifth and sixth graders' understandings of sampling in the context of taking a survey. We did this to develop a sketch of children's early understanding of statistical inference and to examine the effect that everyday contexts have on this understanding and its evolution. We found that contextual statistical settings bring forth ideas not found in the chance setups typically used to teach and test probability understanding. In particular, we found what we labeled a *covariance assumption*. Chance setups using dice and urns involve self-same relations between what is selected to be in the sample (e.g., a blue marble) and what is measured across the population (e.g., blue marbles). However, in many statistical contexts there is a covariant relationship between what is selected to be in the sample (e.g., people) and what is measured across the population (e.g., opinions). In the context of a survey, children tended to focus on the people who generated opinions, judging whether a sample ensured one of each type of person or whether the sampling method was fair to all the people in the population. Further research can illuminate whether these results are peculiar to an opinion survey. Nonetheless, the ability to understand opinion polls is an important mathematical literacy skill. Our results indicated that children do not naturally see the congruency between surveys and chance.

The children brought a host of interpretations to sampling that fell under the umbrella of the covariance assumption. Each interpretation bore a family resemblance to a statistical inference. An invitation, for example, is like a sample in that one selects a portion of the population for inclusion, and a vote is like a sample in that one is trying to find out the distribution of opinions within a population. These family resemblances, however, do not capture the full structure of a statistical inference. Instead, they are more like prototheories in that they provide reasonable initial meaning for drawing inferences about sampling. They need to be reconciled with one another to create a more complete theory of sampling. For example, we found that in the context of a survey the children readily accepted the idea of a stratified sample, but had trouble with a random sample selected by a randomizing device. Yet in the chance context of dice and urns they were able to reason about the effects of a chance device (i.e., a roll of the die), although we doubt that they had a full grasp of probability. Thus, the children had two prototheories of sampling that were in some sense incongruent.

Although we believe that these prototheories provide the basis for understanding the full structure of a statistical inference, they present an instructional challenge. On the one hand, it is necessary to make contact with the students' prior knowledge, and hence we do not believe that one should avoid the contexts that bring these prototheories to mind. On the other hand, because the children's prototheories overlap with the structure of a statistical inference, it may be difficult for children and teachers to assess how a prototheory differs from more normative conceptions of statistics. Moreover, as one tries to accommodate children's prototheories into the fuller structure of a statistical inference, the children may slip among prototheories to understand the different relationships within a statistical inference.

To meet these challenges, we rely on anchored instruction in which multiple prototheories may be juxtaposed within the same context. This is quite different from assigning students problems that extract a single link within the total statistical inference. In this latter approach, students can solve all the problems correctly using an appropriate prototheory without seeing how their understandings for each type of problem may be incongruent with one another. In the final study, with the aid of the SMART assessment model, we helped children reconcile the ideas that a good survey is stratified, whereas a good marble sample is random. We did this through an example in which the survey was simulated by using the chance setup of drawing colored chips that stood for different opinions about a survey. Our results provide some initial evidence that children reconciled their prototheories of stratified and random selection methods into a prototheory of fairness. In this case, their assumption became that a good sample is one in which every type of person has a fair chance of participating in the survey. Although short of a normative understanding of statistical inference (students still focus on the population characteristics more than the target inference), this represents an advancement in their understanding of statistical inference: Stratification and randomness are no longer incongruent.

We think it likely that further contextual instruction, complemented with assessment activities that highlight and bridge specific relationships, will help children build the larger structure of a statistical inference. In addition to helping students align their assumptions about what makes for a good sample, each context can bring new models and examples of sampling. For the uninitiated, drawing names from a hat, or picking people systematically out of line, can seem effortful and arcane. Examples of sampling in action, however, can provide initial credibility. We recall, for example, the instance of college students who were having some trouble randomly assigning 60 subjects to their four experimental conditions. We suggested that they put all the subjects' names into a hat and randomly pull 15 for each condition. Their initial response was, "We thought we had to use a table of random numbers." It was only by providing an alternative model of sampling that we could begin the discussion about why the two methods were analogous and build a deeper understanding of random sampling. Accordingly, in our Jasper Adventure Series and Jasper Challenge Programs, we provide multiple examples of mathematical concepts at play in different contexts, as well as further analogous situations to help children align their multiple understandings of the assumptions at the interface of the everyday world and mathematics.

ACKNOWLEDGMENTS

Additional members of the Cognition and Technology Group at Vanderbilt who contributed to this work include Allison Moore, Linda Zech, Cynthia Mayfield-Stewart, John Bransford, Laura Till, Kadira Belynne-Buvia, Joyce Moore, Tamara Black, Jim Pellegrino, Chuck Czarnik, Taylor Martin, and Linda Barron. This research was supported by a grant from the National Science Foundation (NSF MDR-9252908).

REFERENCES

Bar-Hillel, M. (1980). What features make samples seem representative? *Journal of Experimental Psychology: Human Perception and Performance, 6,* 578–589.

Barron, B., Vye, N. J., Zech, L. Schwartz, D., Bransford, J. D., Goldman, S. R., Pellegrino, J. W., Morris, J. Garrison, S., & Kantor, R. (1995). Creating contexts for community based problem solving: The Jasper challenge series. In C. Hedley, P. Antonacci, & M. Rabinowitz (Eds.), *Thinking and literacy: The mind at work* (pp. 47–71). Hillsdale, NJ: Lawrence Erlbaum Associates.

Bransford, J. D., Franks, J. J., Vye, N. J., & Sherwood, R. D. (1989). New approaches to instruction: Because wisdom can't be told. In S. Vosniadou & A. Ortony (Eds.), *Similarity and analogical reasoning* (pp. 470–497). Cambridge, England: Cambridge University Press.

Bransford, J. D., Sherwood, R. D., Hasselbring, T. S., Kinzer, C. K., & Williams, S. M. (1990). Anchored instruction: Why we need it and how technology can help. In D. Nix & R. Spiro (Eds.), *Cognition, education and multimedia: Exploring ideas in high technology* (pp. 115–141). Hillsdale, NJ: Lawrence Erlbaum Associates.

Brown, J. S., Collins, A., & Duguid, P. (1989). Situated cognition and the culture of learning. *Educational Researcher, 18,* 32–42.

Chi, M., Feltovich, P., & Glaser, R. (1981). Categorization and representations of physics problems by experts and novices. *Cognitive Science, 5,* 121–152.

Cobb, P., Wood, T., Yackel, E., & McNeal, B. (1992). Characteristics of classroom mathematics traditions: An interactional analysis. *American Educational Research Journal, 29,* 573–604.

Cognition and Technology Group at Vanderbilt. (1990). Anchored instruction and its relationship to situated cognition. *Educational Researcher, 19,* 2–10.

Cognition and Technology Group at Vanderbilt. (1992a). The Jasper experiment: An exploration of issues in learning and instructional design. *Educational Technology Research and Development, 40,* 65–80.

Cognition and Technology Group at Vanderbilt. (1992b). The Jasper series as an example of anchored instruction: Theory, program description and assessment data. *Educational Psychologist, 27,* 291–315.

Dean, A. L. (1987). Rules versus cognitive structure as bases for children's performance on probability problems. *Journal of Applied Developmental Psychology, 8,* 463–479.

diSessa, A. A. (1983). Phenomenology and the evolution of intuition. In D. Gentner & A. L. Stevens (Eds.), *Mental models* (pp. 15–33). Hillsdale, NJ: Lawrence Erlbaum Associates.

diSessa, A. A. (1993). Toward an epistemology of physics. *Cognition and Instruction, 10,* 105–225.

Donaldson, M. (1978). *Children's mind.* New York: W. W. Norton.

Fischbein, E. (1975). *The intuitive sources of probabilistic thinking in children.* Boston: Reidel.

Gal, I., Rothschild, K., & Wagner, D. A. (1990, April). *Statistical concepts and statistical reasoning school children: Convergence or divergence?* Paper presented at the Annual Meeting of the American Educational Research Association, Boston.

Garfield, J., & Ahlgren, A. (1988). Difficulties in learning basic concepts in probability and statistics: Implications for research. *Journal of Research in Mathematics Education, 19,* 44–63.

Goldman, S. R. (1985). Inferential reasoning in and about narrative texts. In A. Graesser & J. Black (Eds.), *The psychology of questions* (pp. 247–276). Hillsdale, NJ: Lawrence Erlbaum Associates.

Greeno, J. G., Smith, D. R., & Moore J. L. (1993). Transfer of situated learning. In D. K. Detterman & R. J. Sternberg (Eds.), *Transfer on trial: Intelligence, cognition and information* (pp. 99–167). Norwood, NJ: Ablex.

Heider, F. (1958). *The psychology of interpersonal relations.* New York: John Wiley.

Jacobs, V. (1996). *Children's informal interpretation and evaluation of statistical sampling in surveys.* Unpublished doctoral dissertation, University of Wisconsin, Madison.

Jacobs, V. R., & Lajoie, S. P. (1994, April). *Statistics in middle school: An exploration of students' informal knowledge.* Paper presented at the Annual Meeting of the American Educational Research Association, New Orleans.

Johnson-Laird, P. N., Legrenzi, P., & Legrenzi, M. (1972). Reasoning and a sense of reality. *British Journal of Psychology, 63,* 395–400.

Kahneman, D., Slovic, P., & Tversky, A. (Eds.). (1982). *Judgment under uncertainty: Heuristic and biases.* New York: Cambridge University Press.

Kahneman, D., & Tversky, A. (1972). Subjective probability: A judgment of representativeness. *Cognitive Psychology, 3,* 430–454.

Kelley, H. (1973). The processes of causal attribution. *American Psychologist, 28,* 107–128.

Konold, C. (1989). Informal conceptions of probability. *Cognition and Instruction, 6,* 59–98.

Konold, C., Pollatsek, A., Well, A., Lohmeier, J., & Lipson, A. (1993). Inconsistencies in students' reasoning about probability. *Journal for Research in Mathematics Education, 24,* 392–414.

Kuzmak, S. D., & Gelman, R. (1986). Young children's understanding of random phenomena. *Child Development, 57,* 559–566.

Larkin, J. H., McDermott, J., Simon, D. P., & Simon, H. A. (1980). Expert and novice performance in solving physics problems. *Science, 208,* 1335–1342.

Lave, J., & Wenger, E. (1991). *Situated learning: Legitimate peripheral participation.* Cambridge, England: Cambridge University Press.

Mokros, J., & Russell, S. J. (1995). Children's concepts of average and representativeness. *Journal of Research in Mathematics Education, 26,* 20–39.

Moore, J. L., Lin, X., Schwartz, D. L., Petrosino, A., Hickey, D. T., Campbell, J. O., Hmelo, C., & Cognition and Technology Group at Vanderbilt. (1994). The relationship between situated cognition and anchored instruction: A response to Tripp. *Educational Technology, 34,* 28–32.

Moore, J. L., & Schwartz, D. L. (1994, April). *Visual manipulatives for proportional reasoning.* Paper presented at the Annual Meeting of the American Educational Research Association, New Orleans.

Newman, D., Griffin, P., & Cole, M. (1989). *The construction zone: Working for cognitive change in school.* Cambridge, England: Cambridge University Press.

Nisbett, R. E., Krantz, D., Jepson, C., & Kunda, Z. (1983). The use of statistical heuristics in everyday inductive reasoning. *Psychological Review, 90,* 339–363.

Owens, D. (1992). *Causes and coincidences.* New York: Cambridge University Press.

Piaget, J., & Inhelder, B. (1975). *The origin of the idea of chance in children* (L. Leake, Jr., P. Burrell, & H. D. Fischbein, Trans.). New York: Norton.

Rubin, A., Bruce, B., & Tenney, Y. (1990, April). *Learning about sampling: Trouble at the core of statistics.* Paper presented at the Annual Meeting of the American Educational Research Association, Boston.

Schwartz, D. L. (1995). The emergence of abstract representations in dyad problem solving. *Journal of the Learning Sciences, 4,* 321–354.

Schwartz, D. L., & Goldman, S. R. (1996). Why people are not like marbles in an urn: An effect of context on statistical reasoning. *Applied Cognitive Psychology, 10,* S99–S112.

Shaughnessy, J. M. (1992). Research in probability and statistics: Reflections and directions. In D. A. Grouws (Ed.), *Handbook of research on mathematics teaching and learning* (pp. 465–494). New York: Macmillan.

Spiro, R. J., Feltovich, P. J., Jacobson, M. J., & Coulson, R. L. (1991). Cognitive flexibility, constructivism, and hypertext: Random access instruction for advanced knowledge acquisition in ill-structured domains. *Educational Technology,* May, 24–33.

Tversky, A., & Kahneman, D. (1983). Extensional vs. intuitive reasoning: The conjunction fallacy in probability judgment. *Psychological Review, 90,* 293–315.

Whitehead, A. N. (1929). *The aims of education.* New York: Macmillan.

Yost, P. A., Siegel, A. E., & Andrews, J. M. (1962). Nonverbal probability judgments by young children. *Child Development, 33,* 769–780.

CHAPTER TEN

Assessing Statistical Knowledge as It Relates to Students' Interpretation of Data

Iddo Gal

University of Haifa, Haifa, Israel

The *Curriculum and Evaluation Standards* of the National Council of Teachers of Mathematics (NCTM, 1989) and other sources (e.g., Goldin, 1990) paint a very broad picture of desired mathematical and statistical skills of students, leaving wide latitude to educators in translating general visions into practice. This situation can foster innovation and creativity, yet could also lead to misdirected efforts and lost opportunities for students and for society at large.

In thinking about the goals of statistics education, this chapter distinguishes between two overlapping clusters of desired skills, termed here *generative* and *interpretive*. We argue that the mathematics and statistics education communities have overemphasized the development of generative skills, and that too little attention has been paid to interpretive skills, despite their criticalness. New and presumably improved mathematics textbooks and curricula, although purporting to adhere to the NCTM Standards, still place a premium on procedural skills in statistical topics (e.g., calculating averages, creating graphs), yet allocate little space or time to developing students' ability to make sense of and communicate about the meaning of data.

This chapter aims to explore some conceptual and practical challenges that assessment of interpretive skills poses to teachers and researchers. The focus on assessment issues stems from the observation that the majority of teachers, at least in the United States, have had little opportunity to develop instructional expectations and understanding regarding students' learning processes in statistics (Weiss, 1994). The availability of effective assessments is thus imperative to inform practices of novice teachers and help them build a knowledge base regarding how students learn statistical topics and what knowledge and skills prove difficult to develop.

The chapter is organized in three sections. In the first section I reflect on the goals of statistics education, explore in more detail the nature of interpretive skills that students need to possess, and focus on opinions as a target for instructional and assessment efforts. The second section examines issues involved in eliciting and evaluating students' opinions about data, with illustrations related to students' opinions about data in simple 2-by-2 tables. This section is followed by a discussion of the implications for general assessment practices in statistics education, and for instruction and future research.

INTERPRETIVE SKILLS AND OPINIONS

Any discussion of educational assessment should first consider the curriculum and learning goals that such assessments are expected to serve (Mathematical Sciences Education Board, 1994). Multiple objectives for teaching statistics are presented by various sources (NCTM, 1989; Goldin, 1990; Moore, 1990; American Association for the Advancement of Science, 1993). For the present discussion, we find it useful to group key skills identified by these sources into two interrelated yet distinct clusters, labeled generative and interpretive

Generative Skills. In general, students are expected to be able to generate data, act on data, or otherwise "do" statistics. To acquire generative skills, it is commonly suggested that students gradually master all phases of a statistical inquiry, and develop an ability to design surveys and sampling procedures, gather and organize data, execute needed computations (e.g., averaging), construct graphs and charts, analyze their data and displays, and possibly perform some statistical significance testing.

Interpretive Skills. In general, students are expected to be able to evaluate (and communicate about) the meaning and implications of data. Yet the specific nature of needed interpretive skills will depend on the type of context in which they are needed.

1. *Reporting (active) contexts* are those situations where students are involved in a research project (e.g., study, experiment, survey) and in this context take part in all phases of data collection, manipulation, and analysis, and have to be able to discuss with others their findings and conclusions. Specifically, this requires that students: (a) understand the information contained in, or implied by, statistical indices and tabular or graphical displays

they have generated (and understand the statistical concepts or ideas on which these are based); (b) can reflect on the implications or meaning of such information in light of the questions that motivated their study; (c) are able to describe and explain to others, verbally or in writing, and optionally using tabular or graphical displays, the findings, conclusions, and implications from their project; and (d) are able to reflect about the quality of their data and procedures and exchange views with others about alternative explanations for their findings, given the stated goals of their project.

2. *Listening (passive) contexts* require that students (and adults) be able to make sense of and optionally react to messages with embedded statistical elements they may encounter in, for example, the printed or visual media or in a workplace context. Unlike in reporting contexts, where students take part in all phases of data collection, manipulation, and analysis, in listening contexts students (and adults) do not engage in generating any data or in making any computations. Listeners are not involved with and are thus not necessarily familiar with the process that generated given data, or with the procedures used to analyze the data. Yet they have to comprehend the meaning of any messages that they are presented with, and be both willing and able to critically examine the reasonableness of such messages or claims, or reflect about different implications of any findings being reported.

The demands presented to actors in listening (passive) contexts may differ from those in reporting (active) contexts with regard to at least four interrelated yet separate facets: linguistic attributes and literacy demands of the messages involved, range of statistical topics involved, degree of familiarity with sources for variation and error in the contexts discussed, and degree of need for critical evaluation of a message and its source (Gal, in press). In general, then, the skills students may develop in generative, reporting, and listening contexts may have only a partial overlap.

Opinions as Manifestations of the Interpretive Process

If teachers aim to develop the full range of students' interpretive skills, in light of the preceding discussion, they will have to spread their efforts over a wide terrain of tasks and activities. This may be difficult in light of the current pressures of an ever-expanding curriculum and limitations on teachers' time. One solution may be to focus on problems and on skills that may be common to both reporting and listening contexts. In this context, we observe that, whether students serve as the generators or the receivers of messages about statistical data, a common thread in both cases is that they involve the creation, communication, interpretation, or defense of opinions.

Opinions are the actual end-products of the interpretation process. We want students to develop skills in presenting clear and well-articulated opinions that can be unambiguously understood. Students should gradually develop an ability to present opinions that are sufficiently self-explanatory, that is, that contain enough information about the reasoning or evidence on which they are based so as to enable a person listening to them to judge their reasonableness. The development of students' ability to generate sensible and justifiable opinions (e.g., about the meaning of certain data, or about the validity of arguments that rely on or make reference to statistical data) should thus become a target area for instruction in statistics education.

Very little has been written about issues involved in developing and assessing students' opinions about data. By their very nature, opinions often cannot be characterized as "right" or "wrong," unlike responses to many standard traditional problems in most domains in mathematics education. Instead, opinions are judgments that have to be evaluated in terms of factors such as:

The quality of the reasoning on which they are based

The reasonableness of any arguments presented

The nature of and the relevance of the evidence/data used in their creation or justification

The adequacy of methods employed to generate, process, or analyze the evidence/data

Because opinions are both a central product of the interpretive process and a key gateway into students' reasoning about data, they serve as the focus of this chapter. The range and complexity of the issues involved in eliciting and assessing students' opinions about data is illustrated in the next section with regard to interpretation of information in tables.

OPINIONS ABOUT DATA: THE CASE OF TABLES

Tables can be used merely to store and organize raw data, but often they are used to display summary indices, not only in generative contexts, but also in interpretive contexts that students may encounter in or out of school, such as in textbooks and newspapers, as well as in workplace manuals (Carnevale et al, 1990). When examining tables, students should be able to identify differences between groups, associations between variables, or trends over time, and discuss their or others' opinions about any such trends.

The advantage of analyzing students' interpretation of data in tables is that relatively few data points are involved (compared to, say, a complex graph) and so students' opinions and reasoning strategies can be more clearly studied and discussed. That said, interpretive tasks involving data in tables are quite similar to those involving data presented in other types of visual displays, such as graphs and charts; hence, much of the discussion that follows should equally apply to assessment of students' interpretation of data in other types of visual displays as well.

When asking students to interpret information in tables (and other types of visual displays of data), at least two different types of questions can be posed, termed here *literal reading questions* and *opinion questions*. This distinction corresponds to one first introduced by Curcio (1989) and Wainer (1992) in the context of graph comprehension, although these authors have not extended their application to information in tables, nor have they elaborated on assessment issues, which are the focus of the present chapter.

Literal reading questions require students to "lift" numbers from specific cells in a table (or locations in a graph), or compare two such numbers. In explaining their answers to such questions (see examples that follow), students have to simply point to a data point in a table. Answers to literal reading questions, which are very common in mathematics textbooks, especially in the context of graph reading, are simple and can be unambiguously classified as either "right" or "wrong." In contrast, opinion questions aim to elicit students' ideas about the meaning of the overall pattern of data in a table (or a graph); such questions lead to answers that often cannot be classified as right or wrong. In justifying their opinions, students have to refer to the relationships between several or all of the data elements in the display, and optionally relate numbers to other information or world knowledge they may have to help in ascribing meaning to any patterns they notice.

Issues involved in eliciting and evaluating students' opinions pertaining to information in tables are discussed next, using examples from a research project funded by the National Science Foundation. A central feature of the Statistical Reasoning in the Classroom Project (STARC) was presenting children in elementary and secondary schools with several extended problems, that is, stories about a person (or an organization) who needed to answer a meaningful question based on some data. As part of an individual interview, students were taken through several stages of each problem, and were asked questions about issues such as how to collect data and from whom, what is the meaning of certain data collected, and what conclusions can be drawn from certain displays of data.

One of the contexts in which students were presented with data in tables involved a story about a toy company that plans to market one of two new educational games to children in schools. This context was presented as part of an individual interview to a sample of 20 grade 9 students and 36 grade 12 students from two urban public high schools in the Philadelphia area. Students were ethnically diverse (about 50% minority students), and about 60% in each grade were females. Students were told that the company conducted several surveys to help it identify which of the two games may be more popular, and thus help it decide which of the two games to produce. Following questions about sample size and sampling methods that should be used by the company in carrying out the survey, students were shown results from several surveys that the company presumably conducted, presented in tabular displays as illustrated in Tables 10.1–10.3, and were asked both literal reading and opinion questions.

In this context, some literal reading questions were asked to make sure that children understand the visual representation in a table and to study how children in different ages go about responding to such simple tasks. Regarding Table 10.1, for example, the interviewer explained that the company gave the two games to 50 children, 25 from one school and 25 from another, that after 1 week each child voted on which game she or he preferred, and that the results were tallied in the table. The interviewer then asked, "How many children in school 1 prefer game A?" Virtually all students answered this and similar literal reading questions correctly. Their explanations invariably and correctly referred to the value of the data point in question (which was 20 in this example), which attested that "mechanical" reading of the data in such tabular displays does not constitute a problem.

I next discuss opinion questions, which were described earlier as a key method to explore students' interpretations of data. The next section examines two aspects of elicitation of opinions: how to pose an initial question, and what data values to present to students. A third aspect of the elicitation

TABLE 10.1
Data Shown to Students

	Game A	Game B
School 1	20	5
School 2	15	10

TABLE 10.2
Data Shown to Students

	Game A	Game B
Girls	18	12
Boys	9	11

TABLE 10.3
Data Shown to Students

	Game A	Game B
Boys	18	17
Girls	12	13

process, the questioning that should follow after an initial opinion is obtained, is discussed later, after I examine the evaluation of elicited opinions.

Eliciting Opinions About Patterns in the Data

Question Posing. Tabular or graphical displays can be used to explore whether students can pick out different patterns in the data, as well as their understanding of various statistical ideas and procedures. We have developed the following guidelines for shaping the initial opinion question posed:

1. It should not provide specific hints as to where in the table (or graph) the student should look. (The idea is that the student will have to decide on his or her own what parts of the display to examine, and how different numbers or data points relate to each other; the student's ideas and thinking process about data analysis and interpretation can be thus revealed through follow-up questioning.)

2. It should suggest to the student that a judgment (opinion) is called for, rather than a precise "mathematical" response in the form of a specific number. (We want students to think about more than one possible pattern in the data, not look right away for some numbers to crunch; also, we want students to be sensitive and open to ambiguities or inconsistencies in the data, and not assume that there is a single right answer.)

3. The question should seek an answer that can be of some service to someone operating within the given context; that is, the question should be an outgrowth of a functional "need to know." (Unless students are presented with a reasonably realistic context, they may not use their "real" head and may not bring into the task the full power of their knowledge and reasoning skills; that said, the same situation may not be equally familiar to all students, and they may have different types of world knowledge or make different assumptions about any given context; I discuss this perplexing issue later on.)

An example for a question that implements these guidelines is one presented about the data in Table 10.1, in order to study students' understanding of simple unidirectional patterns. After informing students that they will be asked to explain and justify any opinion they introduce, and alerting them to the need to examine all the information presented and reflect on it before responding, students were asked: "In general, do you think that students from the two schools differ in their preferences, or do they have about the same kinds of preferences for the two games?"

Similar questions were presented about the data in Tables 10.2 and 10.3, to explore students' understanding of bidirectional patterns. For example, after explaining that the toy company conducted another survey to find about the preferences of boys and girls for the games, students were told how many girls and boys were surveyed, and asked: "In your opinion, do boys and girls have more or less the same kinds of preferences, or do they have different preferences?"

In contrast, literal reading questions (such as the question mentioned earlier about Table 10.1, "How many children in school 1 prefer game A?") by their very nature often violate all three principles. They point the student to specific cells in a table. They do not require any interpretation of the meaning of the numbers beyond stating their value, and their phrasing does not invite the formation of an opinion but rather implies the need to provide a single number as an answer. They do not create (as part of, or prior to, presenting the question) an expectation or a sense that the answer given can serve any functional purpose within the given context.

Although the two sample questions just given call for an opinion, they are not completely open-ended and do focus the student on specific aspects of the tables. In piloting table interpretation questions we initially tried a completely free-form phrasing, such as, "What in your opinion can the toy company conclude from the results in this table?" This phrasing seemed to us to be in line with our guidelines for "good" question posing, yet we found that it led to many vague or uninformative responses. For example, students just reiterated the numbers in each cell of the table or the sums of rows or columns of the table, or claimed that "some children liked each game." These responses were not wrong in any (mathematical) sense, yet masked more than revealed students' understanding and reasoning, as was uncovered later in the interview when more specific questions were asked.

We realized that a fully open-ended phrasing can cause some students, especially those in the elementary and middle grades, to not focus on aspects of the table that should be of practical interest to those who conducted the survey and needed to use its results. (This may have been in part due to lack of complete understanding of the context within which data are generated and the ways in data can be put to use, and lack of experience of students in answering open-ended questions of this type; I later return to these issues.) A somewhat more direct form of question posing thus seemed preferred, to narrow down students' visual scanning and analysis of the data in the table, and to reduce the student's need to rely too heavily on contextual knowledge in order to know what to look for.

Choice of Data Values. Different data values can be used in visual displays to help study different facets of students' statistical knowledge and reasoning. The data values chosen can affect the nature and degree of formality of statistical ideas that students have to address when interpreting data. In Project STARC, we were interested in (a) how students compare data coming from groups of unequal sizes, which is akin to comparing results from samples of unequal size, a situation very common in quantitative research, and in (b) how students evaluate the meaning of different magnitudes of differences between groups, an issue that is related to the statistical notion of significance of differences. Tables 10.2 and 10.3 illustrate two examples for a possible combination of the difference in sample size and the magnitude of a difference between groups that can be used in data displays (whether tabular or graphical).

The data in Tables 10.2 and 10.3 are different from the data in Table 10.1 in two key ways. First, the number of boys and girls surveyed is unequal, whereas in Table 10.1 an identical number of students was surveyed in each school. This feature allowed us to see if students can compare groups on a proportional rather than on an absolute number basis. Data values were also chosen so that the magnitude of differences between cells in the tables is a factor to be considered in drawing a conclusion. In Table 10.2, for example, the differences between the number of girls and boys preferring each game are larger than in Table 10.3, where cells differ from each other only by a value of 1.

Overall, students have to notice the differences between cells in a table or other relationships between the numbers; if they do, they have to decide if known differences are worth paying attention to (are significant). Their interpretation may take into account the question that motivated the research, world knowledge activated by the cover story, and the overall size of each group. Note that issues of interpretation of differences or trends come up even though only simple counts are used in the data in Tables 10.1–10.3. Percentages or averages could be used instead, depending on the cover story being presented to students, but then information about sample size will have to be added to the display (which will already include summary statistics); this may make the visual display more crowded and increase the chance that beginning students will miss some important landmarks in the display.

Evaluating Reasonableness of Opinions About Data

When assessing students' computational knowledge in most domains in mathematics, teachers (and sometimes researchers) are often interested in being able to judge quickly and easily whether students' answers make sense (are right) or not (are wrong). Granted, some problems may have one right answer (e.g., how much is 2 + 2?), others may have multiple right answers (e.g., what two numbers, when multiplied, give a product of 24?), and there may be multiple ways of reaching any answer (e.g., the student may apply an algorithm, use trial and error, count on fingers, use blocks, employ certain heuristics, or simply rely on stored knowledge). Yet, in all cases, it can be unambiguously determined if an answer is correct or not, and independently of the reasoning processes or solution strategies that led to it.

For example, there will be no doubt that the answer "12 and 3" is an incorrect response to the question, "What two numbers, when multiplied, give a product of 24?" A teacher does not have to engage in any elaborate questioning process (although this may be advisable) to find how this answer was reached, in order to determine that the answer doesn't make sense. (To be sure, we in no way endorse assessment processes that ignore students' reasoning process. The preceding discussion simply highlights that assessment processes may have multiple goals or levels, each carrying a certain overhead of time and mental effort; a teacher may be able to achieve one goal of assessment, which is very important in a classroom context, although not other goals, by focusing only on correctness of an answer.)

Is it possible to evaluate correctness of an answer apart from the reasoning that led to it when assessing a student's interpretation of data, for example, the student's opinion about the existence or lack thereof of certain patterns or trends in the data? When very simple data displays show a very clear pattern of results, it may be possible to categorize answers as either right and wrong, similarly to answers to computational questions. However, data displays will often contain multiple elements and features and thus can give rise to different types of answers. The question posed about Tables 10.2 and 10.3, for example, asked students to determine if "boys and girls have more or less the same kinds of preferences, or whether they have different preferences." What would be a reasonable answer to this question?

One reasonable answer for Table 10.2 may be, "They have different preferences: Girls prefer A, boys prefer B." At the same time, it is also reasonable to conclude that "Girls like game A and boys equally like both games," depending on whether the difference between 9 and 11 is considered significant, in light of the total (rather small) size of each group, or if a student assumes the difference is a chance result and will not reappear in a different sample. For Table 10.3, a reasonable interpretation may be that, "Boys and girls have the same preferences," as essentially an equal number preferred each game within each gender group. Yet, if a student takes the word "difference" literally, it is not unreasonable to argue that, "More girls liked game A and more boys liked game B." The upshot is that determining the degree of reasonableness may not be a straightforward process as with computational questions, and further, the determination of reasonableness has to take into account the reasoning process and assumptions made by students. A lot hinges on how students interpret seemingly benign terms such as "same" and "different."

To further illustrate the range of issues involved in evaluating the reasonableness of student opinions, I present next abbreviated quotes from students' opinions regarding the data values in Table 10.3. These opinions were all obtained from 12th-grade students (part of the sample described earlier) who previously did not study statistics, although they may have seen data in tabular displays as part of school instruction in mathematics or other subjects. Annotations point the reader's attention to certain issues to be discussed later. Note that these are all initial opinions, that is, those obtained in direct response to the posing of an initial question, although in some cases answers to probes (appearing in brackets) are also listed. Although some answers appear clear, others are more nebulous and illuminate the inherent need for follow-up questioning. Although these answers were obtained in an interview context, I ask the reader to reflect on what questions or problems will come up if the same answers were given as is in a written assessment, where follow-up questioning is not possible.

1. "They like the same games. . . . Because they each were only 1 apart [points to the numbers in each row, 17–18 and 12–13]."

The student gives an unambiguous opinion and justifies it by comparing the relevant pairs of values from each row in the table. The opinion and its justification are reasonable and are presented here as a benchmark against which the remaining quotes can be compared.

2. "Boys and girls prefer different games. The majority of boys [points to 18] prefer A, the majority of the girls [points to 13] prefer B."

The answer is reasonable if one assumes that the student believes *any* difference is a valid difference. Is this assumption justified? The use of "majority" is disturbing, and could cause confusion in reporting contexts.

3. "A larger amount of boys picked both games, but there were 25 girls and 35 boys so of course there will be a larger amount. If you compare 18 to 17 and 12 to 13, that tells me that boys like both games almost the same as girls."

The student verbalizes some of the reasoning process before reaching a conclusion. An explicit statement acknowledges existence of different group sizes and how a seeming conflict between absolute numbers is resolved. (See quotes 7–9 to appreciate this point.) Notice that the final conclusion is a bit ambiguous, in part due to use of "almost the same," but also because of the insinuation that a difference of 1 cannot be ignored.

4. "Game A has more boys that like that game, but game B is almost equal. I saw that there was a ratio of 6 to 4. [What do you mean by ratio?] Six people is definitely a big separation to me [points to the difference between 18 and 12]."

The initial answer (in the first two sentences) is not clear-cut but has multiple elements. It turns out the student compares the numbers in each column, not row. If only the first two sentences were given as an answer (e.g., in a brief open-ended written question), the opinion may be judged unreasonable or "strange," but not enough information would be available to suggest that the student's strategy of using/reading data is incorrect, unless a follow-up question was asked. The use of *ratio* is disturbing as the student is actually referring to a simple difference.

5. "Students who like A and B are the same. [What made you give this answer?] If you add 13 and 17 for program B it equals the same as when you add 12 + 18 for program A. I think you have to do the survey again."

The first sentence presents a clear and seemingly reasonable opinion, which appears quite similar in phrasing to the opinion presented in quote 1. Yet the follow-up question reveals that, unlike in quote 1, this student ignores the thrust of the question, which is about difference between boys and girls, and instead compares overall preferences for games A and B, leading to an unreasonable answer.

6. "Girls prefer different games. Girls don't like game A as the boys. They don't like game B that much either."

An ambiguous answer, which in first hearing sounds confusing in light of the actual data. Follow-up questioning (not shown) revealed that the student compares the absolute numbers in each column. This leads the student to conclude that girls like each game "less" than boys. The student does not provide any indication of being aware of the implications of difference in group size (which was explicitly noted to him when the question and table were initially presented).

7. "Its hard to compare because there are different numbers. It's not a good comparison. You should have an equal amount."

There is no direct answer. The student is aware of difference between number of boys and girls but appears unable to compare preferences without adjusting group sizes. Note the attention to "fairness" of a comparison. (See next quote for comparison.)

8. "There are 10 more boys than girls. If you took away those 10 extra boys it would be equal. Like if you take away 6 from the 18 it would be 12 and if you took away 4 from the 17 it would be 13 and 6 + 4 = 10 so boys and girls agree on the same."

Like the previous answer, the student has trouble with the apparent difference between the number of boys and girls and wants to equate the results. However, the student goes a step further, assumes how the groups will look if the group sizes were equal, and provides what appears like an opinion in the last few words. (See next quote for comparison.)

9. "It might be equal, or not. If you added ten more girls maybe nine would like A and one would like B. Or, it might even them out."

Like the previous quote, the student does not provide an answer but attempts to first resolve the difference in group size. Yet the student realizes that different options exist for how the numbers may play out, so no opinion appears to exist.

10. "Well, in both games, the boys liked it more than the girls did . . . I really don't think girls are into that stuff anyway, so much as boys are . . . I think the girls have totally different tastes than the boys do in the games.

Because boys like all those rowdy games, and the girls like the not so rowdy ones."

The student compares absolute numbers in columns. The student adds comments suggesting that his conclusion is compatible with preconceived assumptions about preferences or behavioral patterns of boys and girls.

11. "Most boys are active, so either one is fine. Girls would probably go for the one that's easier."

The opinion is vague (what does *either one is fine* mean?). Like the previous quote, the student appears to color the judgment with a prior belief, yet, unlike the preceding quote, it does not appear as if the student attempts to link the belief to the data, but the prior belief is expressed at the outset, with no visible direct reference to the actual numbers.

These examples suggest that a wide range of opinions can be expected from students based on the same set of data. As illustrated by the difference between quotes 1, 2, and 3, the same set of data, even when analyzed with the appropriate strategy, can lead to different opinions. The strategy students use to analyze the data—that is, looking at information in rows, in columns, adjusting group size (quotes 8 and 9), and so on—does not necessarily imply what opinion is given, or if an opinion is reasonable. Further, some opinions appear to be based in part on students' prior beliefs or world knowledge (quotes 10 and 11).

Although some answers can be classified right away as more reasonable than others, other answers have ambiguous elements. The upshot is that assessment of the degree of reasonableness of certain opinions, including some that sound ambiguous, but also some that appear very clear on first hearing, cannot be accomplished without also obtaining, through follow-up questioning, additional information about the nature of the evidence a student considers and the way a student reasons about the relationship between elements (rows, columns, cells, points) in the display.

Follow-Up Questioning

Follow-up questioning is needed for a variety of interrelated reasons. In general, when students present an opinion in response to an initial question posed, they naturally focus on describing the outcome of a thinking process ("I think that . . ."). Follow-up questions are therefore required to get students to describe in detail the reasoning process that led to their answer, including the strategy used in analyzing the data, and to help a teacher (or researcher) understand what is the evidential basis for an answer.

A first step is eliciting information about scanning and analysis strategies. A question we used as a starting point is, "How did you decide about your opinion?" In many cases it proved useful to later separately ask (a) what data points were examined ("Can you show me what numbers [things in the graph] did you look at?"), as well as (b) what information was used when reaching an opinion ("So what numbers [things] helped you when you decided on your final opinion?"). Such questions proved essential for determining whether students noticed or examined all parts of a display, whether they missed or did not notice certain pieces of information (i.e., conducted a partial scan), or whether they noticed all parts of a display, yet ignored or otherwise (mis)interpreted given data in a certain way, possibly leading to an inference error. Follow-up questions can further explore what the reasons were for attending to certain parts of a display, and how the overall pattern of data informs the opinion rendered. Assumptions that students make about the context for the problem, and any world knowledge they bring into the situation (see quotes 10 and 11), are brought out in this process as well.

Our emphasis on questions related to scanning and search strategies was motivated in part by findings from earlier work on students' analysis of information in line plots (Gal, Rothschild, & Wagner, 1990). We found that when comparing two plots, many students, especially those in the early grades, pay attention to only some of the data points. Some students, including at the high-school level, may notice several features of a display yet have trouble integrating them into a coherent picture, and so choose to report about only some of the details they actually examined. (Shaklee & Mims, 1982, discuss a related finding about students' attention to only some of the cells in a table when judging covariation.)

More generally, follow-up questioning is required to clarify opinions that are unclear, brief, incomplete, or noncommittal. Such opinions are not uncommon with younger students or with those whose linguistic skills are still developing (including bilingual students). Such opinions may also be more frequent among students who do not anticipate having to justify their answers or who are not used to having to describe their reasoning; this situation may develop in classrooms where discussion is not promoted, or where instruction centers mostly around presentation of brief answers of the correct/incorrect type to computational questions.

A related situation requiring follow-up questioning is that of students who report certain observations or data points they notice but do not provide a summary conclusion, or who use what may appear to a teacher as "noncommitting" phrases (e.g., "*I think they are sort of different a little*"). The use of such phrases is common when students have trouble resolving conflicting pieces of information or relating different aspects of the display to each other, or when they are unsure about the interpretation of certain patterns in the display. The interpretation of noncommitting phrases can present a problem in some cases, and will require careful probing. For example, what is the meaning of "*I think they are sort of different a little*"? Consider the following:

1. Is the quote an attempt to avoid the need to present a clear and committed opinion? (The student is really unsure, or confused, about what the data mean and so uses a "hedging" sentence to not appear unprepared or to avoid embarrassment.)

2. Does the quote represent the students' complete opinion about the insignificance of a difference between certain data points? (The student notices a difference between groups, yet does not consider it worthy of too much attention; this is similar to when a researcher finds a difference between groups that is in the direction of the research hypothesis but that does not reach statistical significance.)

3. Could the student have a firm opinion about the existence of a meaningful difference between groups, yet appear noncommitting because his or her opinion is expressed using linguistic features ("sort of") that comply with modes of speaking common within the student's local culture? (The student actually means "they are different" but colors the opinion with hedging words, similarly to how many contemporary teenagers in the United States tend to sprinkle "like" and "you know" into their utterances.)

An educator cannot be certain that the (lexical) meaning he or she attaches to a student utterance is compatible with the meaning attached to it by the student. There may be confusion between multiple meanings of the same term, or of mathematical and everyday meanings of certain terms (Laborde, 1990). A student may say, for example, the "average here is greater than there," intending the average to be interpreted as the "middle" of a distribution, whereas a teacher thinks of the mathematical mean.

A related issue of particular concern in the context of interpretations of statistical data is the fact that expression of opinions often includes reference to frequencies, probabilities, degrees of confidence, and so forth. As shown in a series of studies by Wallsten and his colleagues (e.g., Wallsten, Budescu, Rapoport, Zwick, & Forsyth, 1986; Wallsten, Fillenbaum, & Cox, 1986), there are wide individual differences in the meaning attached to probabilistic and other related phrases; this meaning is also moderated by the context and the base rate of the phenomenon in question. The upshot is that follow-up questioning has to be considered an inherent part of the critical and ongoing process of negotiation over meaning of technical terms, phrases, or ideas that may appear implicit in students' opinions.

DISCUSSION AND IMPLICATIONS

This chapter introduced the idea that the generation and reporting of opinions is a core outcome of the analysis and interpretation of statistical data. The assessment of students' interpretation of data requires that educators handle multiple challenges inherent in the elicitation, clarification, and evaluation of students' opinions and of the reasoning behind them. As was illustrated, even simple count data in 2 × 2 tables can give rise to a diverse set of interpretive responses, if open-ended questions termed here *opinion questions* are presented.

Based on the same set of data, different students may provide a range of opinions that can all make some sense. However, some opinions will appear unreasonable to varying degrees, perhaps because they are based on partial, incomplete, or erroneous world knowledge. Yet because such opinions constitute what the student knows or believes about the world, they cannot be judged as erroneous until a student's assumptions about the context of the data and about related world issues are fully explored. Opinions may also seem unreasonable because the strategies used to analyze the data were partial or incorrect, as with attention to selected (and possibly incorrect) parts of the display.

Difficulties with interpretation such as those demonstrated in this chapter are not surprising. First, many school children seem to have problems in a variety of areas deemed prerequisite for dealing with statistical tasks, such as with fractions and percents or ratio and proportionality concepts (Kouba, Carpenter, & Swafford, 1989; Behr, Harel, Post, & Lesh, 1992). In addition, as has been repeatedly shown in other areas of mathematics, students who in general can do simple algorithmic problems geared for their age have difficulty dealing with selection and interpretation of data (Garfield & Ahlgren, 1988; Brown & Silver, 1989), with novel problems (English, 1993), with problems requiring mapping of mathematical statements onto everyday contexts and "sense-making" (Silver, Shapiro, & Deutsch, 1993), or with multistep problems (Nesher & Herskovitz, 1994).

A key conclusion is that a credible determination of the reasonableness of students' opinions about data cannot in most cases be made without having access to both the opinion and the arguments on which it is based. Thus, two separate questions or groups of questions must always be presented to students as part of classroom or research-related assessment. The first one should elicit (and clarify) the student's opinion (e.g., "In your opinion, is there a difference . . ."), and the second one should elicit information about the reasons for the opinion and the evidence used to generate or support it (e.g., "What made you come up with this opinion? What data did you consider?").

Based on these two sources of information, where necessary for assessment purposes, a multilevel scoring rubric can be created for each task or question used (see NCTM, 1995). We recommend that scoring rubrics (or a teacher's judgment, if an informal assessment process is conducted) assign the highest scores to responses that:

Are reasonable in light of the given data

Refer to relevant and sufficient evidence

Make correct use of technical terms

Appropriately reference statistical indices derived by the student, or mathematical relationships or patterns the student notices in the data (such as ratios between parts of the display; see Lamon & Lesh, 1992)

Consider (where relevant) issues of variation and reliability of the data

Make sensible assumptions about the source(s) of the data and about the context within which the problem is embedded and in which data originated

The use of assessment formats that elicit only an opinion without a justification or explanation of arguments, primarily multiple-choice questions, will be inadequate in many cases. In addition to the existence of multiple legitimate answers discussed earlier, an inherent danger is that assessors might assume that students who do not give the "right" answer do not understand how to correctly interpret given data, when in fact students may be basing their answers on a different interpretation of the question, on some assumptions external to the assessment situation that are not known to the assessor (discussed later), or on different premises regarding what constitutes a significant or meaningful difference between groups. Such factors may in part be inherent in any classroom response situation; yet if we think of data as "numbers in context" (Moore, 1990), the nature of assumptions students make about a context must be considered.

Opinions about data may depend on or be colored by students' assumptions about events and processes in the(ir) world. An example is students' assumptions about degree of variability or stability over time that can be expected in a certain variable (such as people's heights, food preferences, feelings, social beliefs, etc.). Students' world knowledge may be limited, erroneous, or based on underlying premises that may be questionable or not generally shared (by other students or by the teacher). Students' knowledge, as it is known to the teacher, will also be moderated by the language students can use to communicate their assumptions and thoughts.

The upshot is that a questioning process has to be an integral part of the assessment of students' opinions about data. This is especially true if teachers intend to know more about the logic students use when reasoning about data, and about intermediate solution steps and temporary conclusions students may have reached during the thinking process. Questioning processes, however, are not without limitations. As Ericsson and Simon (1980) pointed out, asking people to describe what they did to solve a certain problem does not provide the same information as asking them to explain the causes of their solution method or conclusions, in part because each type of question places different demands on the person. It has been argued that questions about the content of a prior internal (mental) process elicit answers that are reconstructive and that may be subjected to various types of intentional and unintentional distortions. Thus, the interpretation of students' explanations for their opinions, and of their answers to process-tracing questions, should be made with caution.

Implications for Teaching Practices and for Teachers' Professional Development

Teachers who are unfamiliar with the differences between teaching mathematics and teaching statistics (Gal & Garfield, 1997) may be insensitive to the ease with which one can reach improper conclusions about students' statistical knowledge and interpretive skills when evaluating their opinions about data. A significant effort may be required to train teachers in effective ways to conduct careful and equitable probing of students' answers and reasoning. Even then, it is unrealistic to expect teachers to implement such questioning on a regular basis with more than a few students at a time, due to time pressures and need to attend to and manage multiple activities in the classroom. Furthermore, although a teacher may be able to collect needed information through oral probing in selected cases, we should be concerned about the assessment of students' interpretations of data when responses are obtained through written means, as in project reports and particularly with exams. In such cases, probing is not possible, and students' developing linguistic and communicative skills may become an obstacle.

A partial solution will be found if students reach a stage where they are able, without being cued, to routinely provide full opinions as well as the justification for such opinions. This requires that teachers establish a "culture of explaining" in the classroom. The establishment of the need to be able to explain one's reasoning as a standard demand characteristic of the learning situation can increase the likelihood that students will spend more time reflecting about (and hopefully improving) their opinion before expressing it. Because the stimuli used in opinion questions involve multiple features (e.g., cells of a table, or data points in a graph), students need to be encouraged to examine all of them and spend sufficient time reflecting on their meaning.

Teachers should become acquainted with the unique characteristics of opinion questions and with the features desired in good datasets and data displays, and receive some training in questioning techniques if not already conversant in them. Students have to be coached over time and given feedback regarding the adequacy and clarity of the information they provide. Without a coaching process, it is unlikely that written assessments will enable teachers to fully evaluate the degree of reasonableness of students' opinions. Although these may seem high expectations from teachers and schools, working toward them is essential if we are to achieve the vision embodied in curricular standards in statistics such as those espoused by the National Council of Teachers of Mathematics (1989) and by other organizations.

Over the years, it has been repeatedly argued that students have few opportunities to develop communication skills as part of math or science instruction (e.g., Lappan & Schram, 1989). Further, it is clear that comprehension and progress in the mathematics classroom depend to some extent on linguistic skills (Kane, Byrne, & Hater, 1974), and that many students labeled as minority or bilingual encounter challenges in learning mathematics that are due in part to linguistic differences (Orr, 1987; Mullis, Dossey, Owen, & Phillips, 1991; Secada, 1992). As mathematics teachers become more familiar with the use of performance assessments (Lesh & Lamon, 1992; NCTM, 1995) and with techniques for developing communication skills of students from diverse backgrounds, it may become easier for teachers to also embrace the use of opinion questions and implement questioning processes discussed in this chapter to support the development of students' statistical reasoning and statistical literacy skills.

ACKNOWLEDGMENTS

This chapter is based in part on work funded by a National Science Foundation grant (90-50006) to the University of Pennsylvania. The opinions expressed do not reflect those of the National Science Foundation or the university. I thank Caroline Brayer Ebby and Patricia Mahoney for their assistance in qualitative analysis and in developing coding schemes that informed the writing of this chapter.

REFERENCES

American Association for the Advancement of Science. (1993). *Benchmarks for science literacy*. New York: Oxford University Press.

Behr, M. J., Harel, G., Post, T., & Lesh, R. (1992). Rational number, ratio, and proportion. In D. A. Grouws (Ed.), *Handbook of research on mathematics teaching and learning* (pp. 296–333). New York: Macmillan.

Brown, C. A., & Silver, E. A. (1989). Data organization and interpretation. In M. M. Lindquist (Ed.), *Results from the 4th mathematics assessment of the National Assessment of Educational Progress (NAEP)* (pp. 28–34). Reston, VA: National Council of Teachers of Mathematics.

Carnevale, A. P., Gainer, L. J., & Meltzer, A. S. (1990). *Workplace basics: The essential skills employers want*. San Francisco: Jossey-Bass.

Curcio, F. R. (1989). *Developing graph comprehension: Elementary and middle school activities*. Reston, VA: NCTM.

English, L. (1993). Children's strategies for solving two- and three-dimensional combinatorial problems. *Journal of Research in Mathematics Education, 24,* 255–273.

Ericsson, K. A., & Simon, H. A. (1980). Verbal reports as data. *Psychological Review, 87,* 215–251.

Gal, I., Rothschild, K., & Wagner, D. A. (1990). Which group is better? The development of statistical reasoning in school children. *Teaching Reasoning and Problem Solving, 8,* 3–8.

Gal, I., & Garfield, J. B. (1997). Curricular goals and assessment challenges in statistics education. In I. Gal & J. B. Garfield (Eds.), *The assessment challenge in statistics education* (pp. 1–13). Amsterdam, The Netherlands: IOS Press.

Gal, I. (in press). The numeracy challenge. In I. Gal (Ed.), *Numeracy development: A guide for adult educators.* Cresskill, NJ: Hampton Press.

Garfield, J. B., & Ahlgren, A. (1988). Difficulties in learning basic concepts in probability and statistics: implications for research. *Journal of Research in Mathematics Education, 19*(1), 44–63.

Goldin, J. (1990). *A rationale for teaching probability and statistics in primary and secondary schools: A report of the committee on probability and statistics.* Rutgers University, NJ: Center for Mathematics, Science, and Computer Education.

Kane, R. B., Byrne, M. A., & Hater, M. A. (1974). *Helping children read mathematics.* New York: American Book Company.

Kouba, V. L., Carpenter, T. P., & Swafford, J. O. (1989). Number and operations. In M. M. Lindquist (Ed.), *Results from the 4th mathematics assessment of the National Assessment of Educational Progress (NAEP)* (pp. 64–93). Reston, VA: National Council of Teachers of Mathematics.

Laborde, C. (1990). Language and mathematics. In P. Nesher & J. Kilpatrick (Eds.), *Mathematics and cognition* (pp. 53–69). New York: Cambridge University Press.

Lamon, S. J., & Lesh, R. (1992). Interpreting responses to problems with several levels and types of correct answers. In R. Lesh & S. J. Lamon (Eds.), *Assessment of authentic performance in school mathematics* (pp. 319–342). Washington, DC: American Association for the Advancement of Science.

Lappan, G., & Schram, P. W. (1989). Communication and reasoning: Critical dimensions of sense making in mathematics. In P. R. Trafton & A. P. Shulte (Eds.), *New directions for elementary school mathematics: 1989 yearbook* (pp. 14–30). Reston, VA: NCTM.

Lesh, R., & Lamon, S. J. (Eds.). (1992). *Assessment of authentic performance in school mathematics.* Washington, DC: American Association for the Advancement of Science.

Mathematical Sciences Education Board. (1993). *Measuring what counts: A policy brief.* Washington, DC: National Academy Press.

Moore, D. S. (1990). Uncertainty. In L. A. Steen (Ed.), *On the shoulders of giants: New approaches to numeracy* (pp. 95–137). Mathematical Sciences Education Board. Washington, DC: National Academy Press.

Mullis, I. V. S., Dossey, J. A., Owen, E. H., & Phillips, G. W. (1991). *The state of mathematics achievement: NAEP's 1990 assessment of the nation and the trial assessment of states.* Washington, DC: National Center for Education Statistics, U.S. Department of Education.

National Council of Teachers of Mathematics. (1989). *Curriculum and evaluation standards for school mathematics.* Reston, VA: Author.

National Council of Teachers of Mathematics. (1995). *Assessment standards for school mathematics.* Reston, VA: Author.

Nesher, P., & Hershkovitz, R. (1994). The role of schemes in two-step problems: Analysis and research findings. *Educational Studies in Mathematics, 26,* 1–23.

Orr, E. W. (1987). *Twice as less: Black English and the performance of black students in mathematics and science.* New York: Norton.

Secada, W. G. (1992). Race, ethnicity, social class, language, and achievement in mathematics. In D. A. Grouws (Ed.), *Handbook of research on mathematics teaching and learning* (pp. 623–660). New York: Macmillan.

Shaklee, H., & Mims, M. (1982). Sources of error in judging event covariation. *Journal of Experimental Psychology: Learning, Memory, and Cognition, 8,* 208–224.

Silver, E., Shapiro, L. J., & Deutsch, A. (1993). Sense making and the solution of division problems involving remainders: An examination of middle school students' solution processes and their interpretations of solutions. *Journal of Research in Mathematics Education, 24,* 117–135.

Steen, L. A. (1992). Does everybody need to study algebra? *Mathematics Teacher, 85*(4), 258–260.

Wainer, H. (1992). Understanding graphs and tables. *Educational Researcher, 21*(1), 14–23.

Wallsten, T. S., Budescu, D. V., Rapoport, A., Zwick, R., & Forsyth, B. (1986). Measuring the vague meanings of probability terms. *Journal of Experimental Psychology: General, 115*(4), 348–365.

Wallsten, T. S., Fillenbaum, S., & Cox, J. A. (1986). Base rate effects on the interpretations of probability and frequency expressions. *Journal of Memory and Language, 25,* 571–587.

Weiss, I. R. (1994). *A profile of science and mathematics education in the United States: 1993.* Chapel Hill, NC: Horizon Research.

PART FIVE

EPILOGUE

CHAPTER ELEVEN

Reflections on a Statistics Agenda for K–12

Susanne P. Lajoie
McGill University

Our aim in producing this volume is to shed some light on what is known about approaches to precollege statistics education and assessment. Our premise is that if statistics are introduced in the K–12 period, then students will be better prepared for decision making in the real world. The quest is to find ways to provide students with opportunities to do statistics in a manner that prepares them for real-world experiences. Each of the contributors to this volume has embarked on this quest, and provided insights about how to achieve this goal. One path selected by each contributor is a problem-solving approach to statistics where students are asked to interpret the meaning of statistics as opposed to simply understanding how to compute or generate statistics. Lajoie and Romberg (in the introduction to this volume) describe many forks along this path that provide the reader with ideas about how to pursue an agenda that might pertain to statistical content, teaching, learning, and assessment in K–12. My reflections on these chapters have led me to consider these forks as forming a three-lane highway (the Statistical Content Causeway; see Fig. 11.1) that provides a conduit for establishing an appropriate agenda for statistics education and assessment in K–12. The highway is so named because the goal is to move us toward a successful statistics curriculum where we expect students to learn and when we expect them to learn key concepts. This chapter explores this causeway where signposts for when students are expected to acquire certain skills are provided along the way. The instructional in-roads, learning lanes, and assessment alleys that are instrumental to navigating through this content causeway

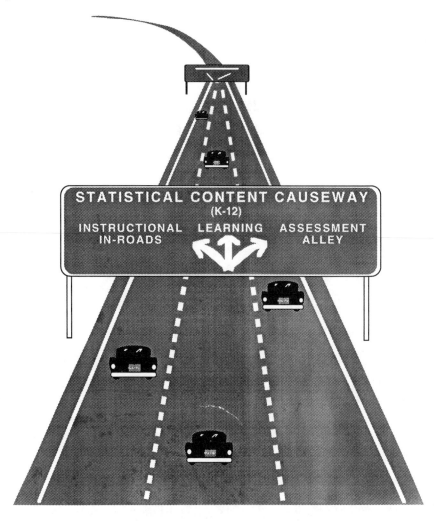

FIG. 11.1. Statistical Content Causeway.

are described. Concluding remarks about possible agendas for statistics education and assessment follow.

THE CONTENT CAUSEWAY

The traffic is heavy along the content causeway for K–12 statistics, and there are many exits and entrances. According to Scheaffer, Watkins, and Landwehr (chapter 1, this volume) students should be intelligent consumers of data and be able to judge the value of data produced by others by the time they reach the end of the causeway, or more specifically grade 12. This is not to say that they should not be able to produce statistics themselves. Having said this, the authors suggest that there are critical periods throughout schooling where certain benchmarks of types of statistical content knowledge should be attained at the completion of high school. The National Council of Teachers of Mathematics (NCTM) *Curriculum and Evaluation Standards* (1989) and the *Assessment Standards* (1995) provide guidelines for selecting statistical content for various grade levels and how one might assess statistical understanding. Scheaffer et al. provide additional details on how statistical curriculum strands can be designed and extended upon through these critical periods. The Scheaffer et al. suggestions are reviewed here in an attempt to see how other contributors have isolated the same content areas as being important areas of investigation and how these authors have contributed to our understanding of the cognitive underpinnings of learning these particular skills and concepts.

Signposts for Critical Learning Periods

Scheaffer et al. describe three critical periods into which statistics strands must be woven: elementary, upper elementary, and high school. Scheaffer et al. pose benchmarks for learning statistics that should be attained during these learning periods. The chapters in this volume address such content and cover the range of K–12 schooling. Scheaffer et al. state that in elementary school, a major benchmark for acquiring statistical understanding is to develop number sense. Linking statistical understanding to number sense can be done by having students collect data and sort, count, and measure items of practical use or interest to them. The significance of developing number sense in elementary school is that once sense-making ability is mastered, students can move on to understanding graphs and tables. Scheaffer et al. suggest that such experiences will help students understand how data are produced in studies planned to answer questions and how to determine when a study is flawed. By the upper elementary grades, experiments and surveys can be used to solidify concepts of number and data and can be repeated in middle school and high school at a more sophisticated level to extend their knowledge. Burrill and Romberg (chapter 2, this volume) suggest that middle school can serve to bridge the gap between informal knowledge about making conjectures from collections of data and the formal ways users of statistics analyze data. Nonetheless, making decisions through data depends on probabilistic reasoning. According to Scheaffer et al., probability as a relative frequency can be introduced in the elementary grades with expansion to notions of probability distributions and expected values in the upper grades. The goal should be for high-school students to develop a full range of analytic skills, from designing a study, through collecting and summarizing data, to making sound inferences based on probability. Scheaffer et al. recommend that students develop the facility for statistical reasoning that allows them to judge the merit of quantitative arguments, rather than being taught inferential techniques per se. The sequencing of statistical content in this manner can facilitate the acquisition of statistical knowledge (Scheaffer et al.). This acquisition process is further described next.

Number Sense and Schema Development. Number sense bears some discussion because it is a precursor to more sophisticated statistical reasoning. Attaching meaning to numbers is critical because to young children numbers are often abstract entities (Resnick, 1988), and methods must be developed to assist students in attaching meaning to these entities. Scheaffer et al. suggest that providing numbers in a context can assist students in making sense of data, especially because numbers can only be compared if they are from the same contexts. As they point out, a 4-pound trout means something different from a 4-pound baby. In one context it is good news, and in another, potentially bad news. The Mathematics in Context project, reported on by Burrill and Romberg, is based on the same premise. Burrill and Romberg have successfully implemented a statistics and probability strand in the middle schools with "context" in mind. They encourage students to understand the role of statistics and probability in their own lives as well as in the context of mathematics. The success of their project is in part due to the abundance of rich and varied problem situations that allow students to develop preformal intuitive notions that lead to more formal quantitiative approaches with more complex problems.

One indicator of the emergence of number sense is that learners become "professional noticers" of data in that they can identify data as being a category, a count, or a measurement. Scheaffer et al. state that students become professional noticers of data by developing appropriate schemas for representing which data goes with which graph. As with Burrill and Romberg, the multiple contexts provide a mechanism for developing more robust schemas, encouraging generalizations whenever needed. Bright and Friel (chapter 3, this volume) provide a more in-depth look at specific types of graphs and how schemas can be developed by providing multiple contexts of which data goes in which graph. Number sense is mapped to measurement when students actively problem solve about data that can be presented graphically. More particularly, links are made between data and variables, and what constitutes a category or frequency, and the selection of graphs based on the data. Graphical representations are an important part of the analysis phase, and Bright and Friel discuss instructional strategies for moving between three different pairs of representations: bar graphs showing ungrouped data and standard bar graphs, line plots and bar graphs, and stem-and-leaf plots and histograms. Their instructional approach emphasizes reading and interpreting such graphs as opposed to making graphical representations. This approach is described next.

Instructional In-Roads. Many of the contributors to this volume describe instructional in-roads for statistics education. A common theme is to build on students' informal knowledge in ways that will formalize statistical understanding. The previous discussion of number sense is a good illustration of how prior knowledge of numbers in context can be used to facilitate the transfer of knowledge of counts to measurements, and finally how such measurements can transfer to graphical representations and the interpretation of patterns in data. In this section, multiple examples are provided of how instruction can be designed to facilitate the extension of knowledge. First, an illustration is presented from Bright and Friel's research that illustrates how students must have an understanding of raw data prior to understanding summary data. Instruction that describes explicit connections between different types of representations is also discussed (Bright & Friel, chapter 3; Friel & Bright, chapter 4; Gal, chapter 10, this volume). Another instructional in-road that holds much promise is the use of scripted questions that provide students with multiple opportunities for providing evidence of their statistical understanding (Gal, chapter 10). Finally, two approaches to extending statistical understanding through multiple contexts are provided, one being the challenge program of Schwartz, Goldman, Vye, Barron, and the Cognition and Technology Group at Vanderbilt (chapter 9, this volume) and the other student-generated statistics projects of Lajoie, Lavigne, Munsie, and Wilkie (chapter 8, this volume).

Bright and Friel have specific recommendations for instruction based on their understanding of the learner. Once again, schema development as it pertains to number sense is critical. Bright and Friel recognize that when students collect data they must represent it, which means making a decision about whether or not to leave the data in its raw form (ungrouped) or to reduce (group) the data in some way. Data reduction can occur by grouping the data values themselves (e.g., number of times data values occur) or by grouping data according to attributes of the elements in the sample (e.g., gender). The decision about grouping has implications for the types of graphical representation selected. Bright and Friel support an instructional sequence whereby ungrouped data precedes grouped data. Their rationale is that students have a better chance of developing a number sense when they can identify with the particular data values in some way. For instance, when dealing with ungrouped data each value has its own plot element such as a bar where students might be able to identify the bar that belongs to them. Once students understand this relationship, they can move on to grouped data. When grouped, the x-axis provides actual data values instead of names of students, and each value has a bar associated with a unit. The

y-axis provides a frequency count of the occurrence of each of the values matched with the heights of each bar.

Bright and Friel discuss tabular representations as well and suggest that a statistic table can serve as an important intervening representation where transitions between raw and reduced data can be made. For instance, using tallies can help students focus on the relationship between data values and their frequencies. The Bright and Friel examples illustrate how students struggle to recognize the connection between mathematical concepts and graphical representation. They also clearly outline how to facilitate statistical understanding by connecting pairs of graphs, and tables and graphs. Contrasting different types of representations in pairs helps students see the necessary relationships between data. Bright and Friel's instructional units are presented with excerpts of interview responses from students using the units. These examples shed light on how students make the necessary connections between representations. The principle behind Bright and Friel's instructional approach is to establish connections or translations among representations to promote understanding. Extending knowledge in this way has been described as an important classroom dimension for facilitating understanding (Carpenter & Lehrer, in preparation). For instance, box plots might be connected to line plots or histograms, and scatterplots might be connected to line plots of the data represented along each axis separately. Bright and Friel dealt with graphical representations, grouped and ungrouped data, and how students discover the relationships between numbers, categories, and graphs. They have developed their instructional activities to facilitate the extension of students' knowledge through sequenced instruction that leads students to understand numerical entities and their concretization through graphs. Multiple examples of each concept are provided to encourage greater flexibility in their knowledge structures.

There is a wonderful symbiosis between Bright and Friel's approach to graphical interpretation and Gal's approach to tabular interpretation. Both support statistical reasoning through multiple contexts and extensions of knowledge. In Gal's Statistical Reasoning in the Classroom project (STARC), designed for the elementary and high-school levels, students are taught to interpret data found in tables. His method was to extend students' comprehension of statistical tables by scripting questions that led students to interpret the data rather than list numbers in cells. His instructional approach entails eliciting student opinions about such data, because he sees opinions as end products of the interpretation process.

Others have examined student opinions about data. Schwartz et al. examined student opinions about sampling as it pertained to fairness and lack of bias. One of their findings was that young students had appropriate schemas for sampling when the procedure dealt with neutral objects, such as selecting different colors of marbles from a hat. However, their sampling strategy deterioriated when asked to sample peoples' opinions about something, such as how people would vote in an election. When opinions were being sampled students thought fair or equal representation was a better sampling method than sample representativeness. Schwartz et al. designed their instructional approach to handle these types of discrepencies in prototheories. They created the challenge program to extend students' statistical understanding by applying their statistical knowledge to new situations. These situations may be crafted to put children's prototheories, such as that of sampling in the two situations just described, in conflict, where social interactions can be used to resolve conflicts.

Another instructional approach was presented by Lajoie et al. in the Authentic Statistics Project (ASP). Their instructional approach was twofold: (a) to engage students in tutorials where basic statistical concepts and procedures were taught through problem-solving activities and (b) to provide opportunities to students to extend their knowledge to self-generated research projects, where they had to design the statistical investigation. Once again the extension of statistical understanding is an important goal of instruction, and in this case the application of statistical knowledge to student generated projects provides a meaningful extension of what was learned in the classroom.

Instructional in-roads for K–12 students have been described in the preceding text. These in-roads have largely been described as working on students' prior knowledge and extending it through multiple examples and multiple connections to new knowledge (chapters by Bright & Friel, Gal, Lajoie et al., and Schwartz et al.). However, because statistics for K–12 is still in its infancy, teachers may need the same learning experiences as students. Friel and Bright (chapter 4, this volume) use complimentary pedagogical principles for teaching teachers and teaching students. In both cases, the learner is encouraged to find meaning in numerical contexts. In addition to extending teachers' understanding of statistical content, teachers' pedagogical beliefs and practice are extended through workshops. Friel and Bright maximize the outgrowth of their workshops by involving their participants in the cycle. For example, once teachers have participated in the workshops they may be invited to be teacher leaders, that is, to teach other teachers about statistics. Hence, teachers extend their own knowledge from that of a learner of statistics to that of a teacher of statistics, where the learner may be other teachers or students. This is very much a cognitive apprenticeship (Collins, Brown, & Newman, 1989) approach for teaching teachers about statistics, because experts shared their expertise with novices, and once novices had demonstrated their understanding, assistance was faded or taken away. After such training, these skilled teachers were asked to model their new skills to novice teachers. Hence, the more skilled teachers were able to assist less skilled teachers acquire the skills to teach statistics to elementary-school children.

In summary, instructional in-roads for K–12 statistics must be designed for both students and teachers, because this is a new domain for both participants.

Learning Lanes. As seen earlier, extending students' knowledge of statistical content through multiple examples can provide a fast track in the learning lane. However, in addition to the development of statistical content and specific activities for these statistical strands, active research is needed to identify the cognitive underpinnings of statistical knowledge in K–12. Many researchers are trying to identify how students learn statistics and what the "learning lanes" or transition paths are to developing statistical competence.

In this volume, three chapters specifically examine the development of children's intuitions regarding models of chance, samples, and probability: Horvath and Lehrer (chapter 5), Metz (chapter 6), and Schwartz et al. (chapter 9). The chapters by Horvath and Lehrer (second graders, fourth/fifth graders, adults) and Metz (primary grades) carefully examine the dimensions of students' intuitions and the transitions in this knowledge that reflect emerging statistical competence. Horvath and Lehrer found that children (second graders) did not understand the role of sample space without significant instructional support, be it conversational facilitation or notational supports. They found that fourth/fifth graders needed less assistance to reason about uncertain phenomena and were able to apply their models of uncertain phenomena to new tasks. As expected, adults were more expert still. Horvath and Lehrer and Metz provide insight regarding the informal knowledge children have about probability that can be used to build more formal models of how their statistical understanding develops.

Schwartz et al. examine fifth and sixth graders' informal knowledge regarding sampling assumptions and specifically address the notion that children have statistical prototheories about chance and samples in the context of surveys. Schwartz et al. examine children's models of sampling and chance as a first step in developing instructional challenges that will stimulate the growth of these models through the confrontation of conflicting prototheories. In particular, they describe the development of statistical knowledge as akin to diSessa's (1983, 1993) view that knowledge evolves in pieces of intuitive understanding that are dependent on the problem context. Rather than examine misconceptions, Schwartz et al. suggest looking at the normative development of statistical reasoning. They suggest that people do not necessarily have misconceptions about probabilities, but rather have competing understandings. Knowledge relevant to answering a question about a probable event is differentially applied depending on the pocket of understanding that is drawn on for a particular situation. Rather than beginning instruction on abstract principles (i.e., conjective probabilities), instruction should include situations in which competing intuitions can emerge and be reconciled into a more robust body of understanding.

Different contexts can invoke different types of learning. Gal is particularly interested in the different uses of generative and interpretive contexts for statistics education. His contention is that generative contexts, where students are asked to do statistics, to graph, to compute, and to analyze data, are used far more frequently than interpretive contexts where students are asked to make sense of data to evaluate statistics in some manner. He specifically addresses ways of instructing and assessing statistical understanding of tabular data in interpretive contexts. He argues that whether or not students generate or receive statistical messages, they still have to form and defend their opinions as to what these messages mean. The validity of these arguments should be the focus of instruction and assessment.

These types of contexts are discussed by other contributors, although by other names. For instance, Scheaffer et al. describe contexts where students are asked to be producers or consumers of statistics. Others have described this dichotomy as producers or critiquers (Palinscar & Brown, 1984). Acquiring a critical attitude is central to the Mathematics in Context project described by Burrill and Romberg. Students learn to raise questions in a problem-solving context, attempt a hypothesis and ways to answer the questions, specify variables and their interrelationships, and collect data on these variables. Students are asked to build a case to answer questions based on their information and analysis. Their approach is similar to those described by Lajoie et al., Schwartz et al., and Derry, Levin, Osana, and Jones (chapter 7). For example, Lajoie et al. used technology as a springboard for demonstrating different levels of student performance as it applied to generating statistical projects. Consequently, the student-generated projects that were modeled served as a context for the process of statistical investigation. Schwartz et al. used technology as well. A videodisc was created that provided a macrocontext for statistical problem solving. This technology was used to present extended problems that served to elicit and anchor children's multiple intuitions re. statistical inference. Derry et al. used film as a medium or context for learning to reason statistically. The film *Lorenzo's Oil* was used as an instructional tool, as a model of statistical argumentation, and then as a stepping stone to simulation games where students develop their own statistical arguments using evidence to defend their positions. Derry et al. created situations where students took on various roles in a debate—supporter, opponent, or judge—thereby making the learning context different for each role.

To summarize, the contributors to this volume have described a variety of learning lanes. One common ingredient in paving these lanes is that of building on prior knowledge and extending statistical understanding through multiple examples. Horvath and Lehrer have started to identify the differences in statistical competence as it pertains to concepts of chance, for different age levels. Metz compliments the Horvath and Lehrer chapter in that similar dimensions of understanding chance were purported. Schwartz et al. suggest ways to stimulate the transitions between such levels of understanding, through direct confrontation of student prototheories by the presentation of competing prototheories. These chapters taken together provide the cognitive underpinnings for the development of theories of chance and sampling. We also have provided evidence that different instructional contexts result in different types of learning (chapters by Burrill & Romberg, Derry et al., Gal, Lajoie et al., Scheaffer et al., and Schwartz et al.).

Assessment Alleys. Although three chapters explicitly address how to integrate assessment with instruction (Gal, Lajoie et al., and Schwartz et al.), other chapters also address assessment. Lajoie et al. and Schwartz et al. describe ways in which technology can be used to enhance the assessment process through modeling student performance so that students can internalize the assessment goals of instruction. Gal describes a multilayered approach to assessment through the use of scripted questions. His rubrics for scoring the reasonableness of student opinions are quite useful and have some interesting overlap with the work of Derry et al. on examining the reasonableness of students' statistical arguments. These assessment alleys are described next.

In a middle-school study, Lajoie et al. developed assessment models of statistical performance that were integrated into the instructional sequence. Students were taught about the statistical process of investigation through exemplars or performance benchmarks of statistical competence. Technology supported this modeling process by generating video images of students performing statistical investigations. These assessment models were utilized by students in establishing their own expectations for success (Lavigne & Lajoie, 1996). The Lajoie et al. chapter demonstrates the use of multiple assessment methods where students had multiple opportunities to give evidence of their statistical understanding. Evidence of student understanding was found in pre- and posttest data, journals, verbal protocols collected during observation of student models in the authentic stack, verbal protocols during preparation of statistics projects, the statistics project presentations themselves, and through homework assignments. Technology supported assessment in two ways. First, it was used to model the instructional goals to students, as well as making assessment criteria clear and open to learners. Second, technology was used to support scoring the student data, when data were generated using computers. A scoring template was developed using a hypercard to demonstrate examples of students using the computer to develop their statistics project. The template gave concrete examples of what the scoring rubrics entailed.

Schwartz et al. provide an integrated approach to statistics instruction and assessment by supporting a tandem process of discrimination and reconciliation of statistical prototheories where both informal and formal mathematical understanding were confronted by instructional challenges. By assessing understanding in this instructional context, developmental transitions could be observed. The instructional challenges are designed to encourage student reflection by graphical representations of how other students solved similar statistical problems. These reflection opportunities are great for promoting small-group discussions where students can critique the presenters' thinking and negotiate meaningful suggestions. Hence, students become more aware of the instructional and assessment goals.

In the challenge for the next time portion of the work of Schwartz et al., students view program segments and are asked to imagine a new sample size and how a similar study might be conducted. Thus, students are asked to apply their knowledge to new situations. Students write their responses, which are evaluated by teachers who provide written feedback. The assessment of student responses is facilitated by technology. The student answers can be easily scored by presenting teachers with a computerized checklist of what forms of feedback are most appropriate for what kind of student response. Quantifying the feedback process to qualitative responses can ease the assessment burden for the individual teacher.

A multilayered approach to assessment is supported by Gal's research in that more than one source of evidence is used as an indicator of student understanding of data. STARC is presented to children in elementary and secondary schools. Several extended problems are presented and students must answer a meaningful question about the data. Gal concentrates on establishing appropriate assessments of the reasonableness of student opinions. He states that when students defend their opinions they should provide enough information about their reasoning and evidence so that a person listening to the argument can judge the reasonableness of the argument. Consequently, he has designed opinion questions into his instructional units. The opinion questions address the meaning of the overall pattern of the data, rather than the correctness of the data. Students are interviewed about the stages in their problem solving, how to collect data, from whom, the meaning of data collected, and what conclusions could be drawn from certain tabular displays of data. The degree of reasonableness of their opinions is followed up by questions that provide more information about the nature of evidence that students consider and how they reason about the elements in a display.

Scripted questions, such as those described by Gal, are a valid assessment option for testing students' statistical reasoning. However, there still may be difficulties in interpreting student utterances because there may be multiple meanings or confusions in mathematical and everyday terms (Laborde, 1990) that may be context dependent. Gal suggests that assessment formats that elicit an opinion without a justification will be inadequate because they assume that an incorrect answer reflects that a student did not know how to interpret the data, when in fact, it could be that the student did not represent the question the same way others did, or is basing his or her answer on assumptions that are external to the problem. Without appropriate questions, students will not be assessed properly. Gal suggests that two separate questions should be presented: one to elicit and clarify the student's opinion, and the second to elicit and probe for information about the reasons for the opinion and the evidence used to generate or support it. These questions are designed to illuminate students' logic in reasoning with data and their intermediate solution steps and temporary conclusions during the thinking process. Gal developed a multilevel scoring rubric to address reasonableness in light of given data, references to relevant and sufficient evidence, correct use of technical terms, appropriate reference to statistical indices, issues of variation and reliability of the data, and making sensible assumptions about the source of the data and about the context in which the problem is embedded and in which the data originated. These scoring rubrics help to diagnose student difficulties.

Just as Gal examined student opinions about tabular representations to examine the depth of their statistical understanding, Derry et al. examined grade 8 student arguments in statistical gaming situations to see whether or not students could provide reasonable evidence to support their positions. Derry et al. also provided complex scoring rubrics to assess statistical reasoning. Derry et al. concur with Gal that statistical reasoning does not necessitate computation or generating statistics, but always involves interpreting and reasoning about real-world problems with conceptual structures representing such ideas as probability, correlation, and experimental control. The Derry et al. scoring scales took these conceptual structures into account and hence examined student arguments in terms of evidence of correlational reasoning, citing of specific counterpositive evidence, recognizing the need for large sample sizes and/or chance versus real relationships, recognizing that correlation need not imply causation, taking into account methodological quality of the research, noting further research needed before conclusions can be drawn, and categories for inappropriate deterministic thinking and unsubstantiated opinions.

As just demonstrated, assessment of statistical understanding can take on many forms, but the chapters described are aligned with the NCTM *Assessment Standards* (1995). One of the major purposes of assessment is to monitor student progress, and this was described in detail by Lajoie et al. Not only do teachers monitor student progress, but students learn to reflect on their own progress. The openness standard is designed to promote students' self-reflection by making assessment goals clear to learners. In the Lajoie et al. study an example was provided of how performance standards were made inspectable by learners and how learners came to internalize these goals and assess their own progress. Schwartz et al. also developed assessments that foster student reflection and communication about statistical reasoning, by using graphical representations of student work. One of the key messages in the assessment standards is that assessments must provide valid indicators of what students understand. Gal has maximized the validity of his data though the multilevel nature of assessment. He developed interviews, scripted questions, and follow-up questions within a statistical problem-solving activity involving sense-making of data found in tables. His questioning techniques allow for multiple opportunities and types of ways of extracting student reasoning so that reliable and valid indicators of their statistical knowledge are presented. Derry et al. also developed complex scoring scales to examine the complexity and validity of student arguments.

CONCLUSION

As researchers, our ultimate goal is to build a model of what statistical competence means in different content areas and at different grade levels. Such a model can provide some benchmarks for teachers and students alike so that they know what to expect and strive for at different times in the process of acquiring statistical understanding. This model is emerging in this volume, but there is much work to be done. What we do have is a better understanding of what the content causeway for statistics looks like and how to traverse the causeway through the instructional roads, learning lanes, and assessment alleys. Signposts for critical learning periods—elementary, upper elementary, and high school—have emerged, thanks to the work of Scheaffer et al. and Burrill and Romberg. Examples of students' statistical understanding in K–12 classrooms have been provided. The goal should be for high-school students to develop a full range of analytic skills, from designing a study, through collecting and summarizing data, to making sound inferences based on probability. The facility for statistical reasoning should be developed, allowing students to judge the merit of quantitative arguments. If at all possible, Scheaffer et al. encourage teachers to use and teach statistics throughout the entire curriculum, primarily in mathematics and science but with support and application in the social sciences, health, and other academic subjects. Derry et al. demonstrated a mechanism for supporting this interdisciplinary connection.

Classrooms that provide opportunities for rich discourse are needed to focus students on the similarities and differences between representations highlighting the important statistical features. Bright and Friel advise teachers to create opportunities whereby students can compare multiple representations of the same data set, and to establish what Gal calls a culture of explaining within the classroom. There is consensus on the pedagogical front that statistical content should be problem and project based, should build on students' informal knowledge, can be facilitated by a community of learners, and should provide multiple and realistic or meaningful (authentic) contexts. As Lajoie et al. and Schwartz et al. demonstrated, technology can be used in the classroom to facilitate project-based statistics contexts.

Technology is being used as an instructional facilitator as well as an aid to assessment in some situations. These issues are addressed more completely in chapters by Lajoie et al. and Schwartz et al., who use technology in both instruction and assessment of statistical understanding. Technology can be incorporated in ways that demonstrate multiple linked representations simultaneously (as demonstrated by Hancock, Kaput, & Goldsmith, 1992, and in chapters by Lajoie et al., Schwartz et al., and Bright & Friel). For example, multiple graphical representations of the same data could appear on the screen at the same time so that individuals could choose which representation was more appropriate for the data at hand. Technology can change the ways in which data are represented as well as what questions are asked of the data (Bright & Friel, chapter 3).

We have demonstrated that the agenda for K–12 statistics is still open for discussion, but the groundwork for the content causeway has been done. Instructional roads, learning lanes, and assessment alleys have begun to be tested. Educational practice and empirical research will tell us whether or not this causeway will be well traveled. One thing that strikes me about this volume is the number of linkages between content, instruction, learning, and assessment. It is quite possible that a better model of statistical competence could surface if we took a common theoretical position that would link these areas more effectively. Because statistics is interdisciplinary, the approaches to studying statistics include the diverse perspectives held by statisticians, mathematical educators, psychologists, and educators. One common denominator held by the authors in this book is that they all approach statistics as a problem-solving task.

Marshall (1995) described a position on problem solving that unifies content, instruction, learning, and assessment. Although her examples are situated in arithmetic story problems, her theory is broad enough to apply to other domains. Her theoretical focus is on schema development in problem solving. Perhaps her model could help the statistics education and assessment field as well. Marshall articulated her theoretical position on schema theory, how to use what we know about schemas to improve people's learning, how schemas can be used to direct the assessment of learning, how understanding schemas can lead to better understanding of memory, and how we can use this understanding to create satisfactory models of learning and performance. If this approach was used to identify the appropriate schemas in specific statistics content areas, then instructional design and the assessment of transitions in learning could be facilitated as indicated by her theory of schema development. Many of the authors contributing to this volume have already started identifying the appropriate schemas at specific critical periods (Burrill & Romberg, Scheaffer et al.) and for specific content, such as sampling and the notions of chance and uncertainty (Horvath & Lehrer, Metz, Schwartz et al.) as well as for broader issues of graph and table interpretation (Bright & Friel, Gal), and specific statistical reasoning skills that correspond to generating statistical projects (Lajoie et al.) or formulating statistical opinions and arguments (Derry et al., Gal, Schwartz et al.). Furthermore, each of these authors has proposed methods for extending schemas through appropriate forms of instruction, and assessing statistical understanding through appropriate methods. What Marshall could offer is a possible standardization of certain techniques that could assist us in modeling statistical understanding.

To conclude, this volume provides ample suggestions for statistics education and assessment in K–12, but it is evident that it may take several years to find the balance between the breadth and depth of statistical content needed at various grade levels. Although the statistics agenda is still open for exploration, different avenues have been explored in this volume, and only experience and research can further inform us regarding the appropriateness of certain theoretical positions and selected methodologies. Only by traveling these new avenues will we discover whether or not they are unimpeded or peppered with potholes.

Next Steps

Given that we have made some progress in our understanding of precollege instruction and assessment, I would like to boldly suggest some conceivable next steps for educators and researchers to pursue. These suggestions are gleaned from the research described in this volume. In an ideal world, a program of research might appear as follows:

- Select a theoretical paradigm as a foundation to the research that will support your views on learning, teaching, and assessment.
- Form partnerships between teachers, content matter experts, curriculum developers, computer experts, and researchers that are interested in the development of effective forms of statistical instruction at specific grade levels.
- Establish pilot programs that ensure the necessary contact time with students and the necessary evaluation of materials to see whether they are effective tasks.
- Select a specific statistical content area for a specific grade as a context for research and then design studies that identify the structure and evolution of such knowledge. For instance, some solid propositions were put forth by Scheaffer and others that suggest that detailed research

on such content as number sense and its relationship to graphical and tabular representations may assist educators in forming the necessary links for students between simple number concepts and the measurement of data. What would a model of statistical thinking look like in a specific content area, and how could scoring rubrics be designed to document changes and transitions in such thinking? A healthy research avenue, then, is to identify the cognitive processes that underlie statistical problem solving in such contexts. Identification of such skills within specific content areas can lead to the design of effective instructional materials that will encourage the necessary transitions in learning. Assessment tasks will need to be developed along with such instruction so that valid evidence of learning can be acquired in instructional contexts.

- Build on the cognitive literature that already exists to further develop a conceptual model of statistical problem solving. For instance, build on the schema development models to design studies that will result in a better understanding of the transitions between number sense and data representation as used in graphs and tables. Examine the transitions between planning and performing statistical investigations, generating and interpreting statistics, or producing and critiquing statistics.

- Examine and document ways of transferring learning from one statistical context to another. One fruitful avenue appears to be the use of multiple examples and statistical situations so that knowledge is not bounded to one type of representation.

- Research on the effectiveness of instruction with and without technology should be done in tandem where results from both avenues inform each other. Further research on how technologies can facilitate the "doing of statistics" and the assessing of statistical problem solving must be conducted. Lajoie, Jacobs, and Lavigne (1995) reviewed several exemplars of how technology can be used to empower precollege students in the use of statistics (Hancock et al., 1992; Konold, Pollatsek, Well, Lohmeier, & Lipson, 1993; Lehrer & Romberg, 1996; Rosebery & Rubin, 1989). These studies clearly point to how technology can lead students to better identify with what the data mean in both a graphical and analytical sense. Students start to identify with data points and learn, often through visualization and animations, what distributions, variation, and central tendency really mean. The International Association for Statistical Education 1996 Round Table Conference on Research on the Role of Technology in Teaching and Learning Statistics has produced a volume of conference papers that should be of interest to this audience. The range in uses of technology are quite vast but not exhaustive, from graphic calculators to uses of the Internet for faciliating useful inquiries regarding chance.

Technologies for teaching and assessing understanding of statistics facilitate student understanding by (a) providing active interactive learning opportunities where multiple representations, and dynamic statistical notations (Kaput, 1992), can be used to extend one's comprehension of statistical data, graphs, and analysis; (b) providing opportunities for self-assessment and promoting metacognitive strategies about the statistical investigation process (Lajoie et al., chapter 8, this volume); and (c) helping students confront misconceptions by through simulations (Nickerson, 1995) or direct challenges (Schwartz et al., chapter 9, this volume). A useful agenda for those using technology for instruction and assessment purposes is to follow up on some of these positive avenues and build on the foundational work established by those not using technology, such as Bright and Friel's research on graphical representations, and Gal's work regarding tabular understanding. This agenda would also allow us to see how the introduction of technology may facilate the learning and assessment process to expand on the efficiency of transfer of knowledge through dynamic representations and multiple contexts for acquiring statistical knowledge. Further, the research done by individuals using technology can also feed back to educators who do not use technology. For instance, self-assessment and metacognitive strategies could still be encouraged in noncomputer environments through the development of examples of student work that demonstrate different levels of statistical reasoning. Models of statistical thinking and reflection could be encouraged through examples using "live" or videotaped students performing similar tasks. Simply observing something happen, whether it be on a computer or TV screen, or through a live performance, will not guarantee that the learner has made the connection between, say, a graphical representation and the meaning behind the representation. It is often necessary to make the direct connection for the learner.

When such examples are made available to students, learners can better attempt to meet the teachers' expectations, thus facilitating both instructional and assessment goals. As new assessment techniques are being developed for statistics, it is necessary to consider how technology can help teachers manage the assessment process. Just as students will need time to adapt to new technologies in the context of instruction, teachers will need assistance in learning how such technologies can be used to facilitate instruction and assessment.

ACKNOWLEDGMENTS

Preparation of this chapter was made possible through funding from the Office of Educational Research and Improvement, National Center for Research in Mathematical Sciences Education (NCRMSE). I gratefully acknowledge the assistance of Thomas Romberg and the NCRMSE Statistics Working Group for their help in formulating many of these ideas. Thanks to Deborah Metchette for her graphical contribution (Fig. 11.1).

REFERENCES

Carpenter, T. P., & Lehrer, R. (in preparation). Learning mathematics with understanding. In E. Fennema & T. Romberg (Eds.), *Classrooms that promote understanding*.

Collins, A., Brown, J. S., & Newman, S. E. (1989). Cognitive apprenticeship: Teaching the craft of reading, writing, and mathematics. In L. B. Resnick (Ed.), *Knowing, learning, and instruction: Essays in honor of Robert Glaser* (pp. 453–494). Hillsdale, NJ: Lawrence Erlbaum Associates.

diSessa, A. A. (1983). Phenomenology and the evolution of intuition. In D. Gentner & A. L. Stevens (Eds.), *Mental models* (pp. 15–33). Hillsdale, NJ: Lawrence Erlbaum Associates.

diSessa, A. A. (1993). Toward an epistemology of physics. *Cognition and Instruction, 10,* 105–225.

Hancock, C., Kaput, J. J., & Goldsmith, L. T. (1992). Authentic inquiry with data: Critical barriers to classroom implementation. *Educational Psychologist, 27,* 337–364.

Kaput, J. J. (1992). Technology and mathematics education. In D. A. Grouws (Ed.), *Handbook of research on mathematics teaching and learning* (pp. 515–556). New York: Macmillan.

Konold, C., Pollatsek, A., Well, A., Lohmeier, J., & Lipson, A. (1993). Inconsistencies in students' reasoning about probability. *Journal for Research in Mathematics Education, 24*(5), 392–414.

Laborde, C. (1990). Language and mathematics. In P. Nesher & J. Kilpatrick (Eds.), *Mathematics and cognition* (pp. 53–69). New York: Cambridge University Press.

Lajoie, S. P., Jacobs, V. R., & Lavigne, N. C. (1995). Empowering children in the use of statistics. *Journal of Mathematical Behavior, 14*(4), 401–425.

Lavigne, N. C., & Lajoie, S. P. (1996). Communicating performance standards to students through technology. *Mathematics Teacher, 89*(1), 66–69.

Lehrer, R., & Romberg, T. A. (1996). Exploring children's data modeling. *Cognition and Instruction, 14*(1), 69–108.

Marshall, S. P. (1995). *Schemas in problem solving.* New York: Cambridge University Press.

National Council of Teachers of Mathematics Commission on Standards for School Mathematics. (1989). *Curriculum and evaluation standards for school mathematics.* Reston, VA: Author.

National Council of Teachers of Mathematics Commission on Standards for School Mathematics. (1995). *Assessment standards.* Reston, VA: Author.

Nickerson, R. S. (1995). Can technology help teach for understanding? In D. N. Perkins, J. L. Schwartz, M. Maxwell West, & M. Stone Wiske (Eds.), *Software goes to school: Teaching for understanding with new technologies* (pp. 7–22). New York: Oxford University Press.

Palinscar, A. S., & Brown, A. (1984). Reciprocal teaching of comprehension-fostering and comprehension monitoring activities. *Cognition and Instruction, 1,* 117–175.

Resnick, L. (1988). Teaching mathematics as an ill-structured discipline. In R. Charles & E. A. Silver (Eds.), *The teaching and assessing of mathematical problem solving* (pp. 32–60). Reston, VA: National Council of Teachers of Mathematics.

Rosebery, A. S., & Rubin, A. (1989). Reasoning under uncertainty: Developing statistical reasoning. *Journal of Mathematical Behavior, 8,* 205–219.

Author Index

Subject Index

generation guidelines for teaching statistics in K–12, xiv, xvi
inferential reasoning, 24

I

Incertitude/indeterminacy, analysis of children's intuitions, 158–160
Incongruency, sampling assumptions, 236–237, 258, 267, 270
Indeterminacy, *see* Incertitude/indeterminacy
Inert knowledge, mathematics learning, 236
Inferential reasoning
 data analysis, 14
 high-school graduate understanding of statistics, 5, 21–24
 line plot versus bar graph representations, 79
 links in Toolbox segment and sampling issues, 261
 Mathematics in Context curriculum, 42
Information processing
 bar graph
 grouped and ungrouped data, 69, 72–73
 versus line plot, 79
 stem-to-leaf plot versus histogram, 80
Insights into Data unit, Mathematics in Context curriculum, 46, 48
Instruction
 aligning sampling prototheories, 258
 analysis and chance/probability incorporation into mathematical curricula, 169–171
 anchored, *see* Anchored instruction model
 assessment goals in statistical learning, 308
 bar graph of ungrouped versus grouped data, 69–74
 design and statistical learning, 303–306
 extending techniques of graphical representations, 82, 85–87
 interpretative skills, 275
 line plot versus bar graph, 74–79
 statistics/probability activities in primary grades, 154–156
 stem-to-leaf plot versus histogram, 80, 83–84
Interpretative skills
 context and development, 276–278
 generative skills comparison and statistical competency, 307
 goal of statistics education, 275
 graphical representations, 82, 85
 professional developmental program in Teach-Stat, 94
Interval, histograms, 80
Intervention

sampling assumptions and anchored instruction, 263
 statistical reasoning in middle school, 181–182, 191
 teaching statistics in K–12, xv
Interviews
 interpretation of table data, 279, 285–287
 responses
 bar graphs, 69–74
 line plot versus bar graph, 75–79
 stem-to-leaf plot versus histogram, 80, 83–84
 teacher and staff development program of Teach-Stat, 103–104
Intuition
 analysis of children's
 expected distributions of outcomes, 165–169
 incertitude and indeterminacy, 158–160
 likelihood of events, 160–165
 part-whole relations, 157–158
 relative magnitude, 156–157
 development
 elementary school, 302
 knowledge transition, 306
 teaching probability, 18–19
 randomness and understanding, 150
 structuring in primary grades, 169–170

J

Job, student and data analysis of studying habits, 13–14
John's trial, statistical reasoning, 180–189
Joint information, high-school graduate understanding of statistics, 7
Journal, structured and monitoring student progress, 208, 219

K

K–12, reasons for teaching statistics, xii
Kids Online, sampling assumptions and anchored instruction, 261–262
Kindergarteners, knowledge of relative magnitude, 156–157, *see also* Primary grades
Knowledge
 acquisition and monitoring student progress, 209–213
 assessment
 data interpretation skills and opinions, 276–278
 discussion and implications, 289–293
 opinions about data: table cases, 278–289, *see also* Tables
Knowledge of pieces, goal of instruction, 234, 257

L

M